Transit-Oriented Development

Ren Thomas • Luca Bertolini

Transit-Oriented Development

Learning from International Case Studies

palgrave
macmillan

Ren Thomas
School of Planning
Dalhousie University
Halifax, NS, Canada

Luca Bertolini
Department of Geography, Planning,
and International Development Studies
University of Amsterdam
Amsterdam, The Netherlands

ISBN 978-3-030-48469-9 ISBN 978-3-030-48470-5 (eBook)
https://doi.org/10.1007/978-3-030-48470-5

This Palgrave Pivot imprint is published by the registered company Springer Nature Switzerland AG.
The registered company address is: Gewerbestrasse 11, 6330 Cham, Switzerland

Contents

List of Figures

CHAPTER 1

Introduction to Transit-Oriented Development

Abstract Transit-oriented development has quickly become a popular concept: it is a complex process and strategy that can contribute to more sustainable transportation patterns, decreased emissions, and enhanced regional connectivity. A significant percentage of households would like to live in high-density housing near transit. However, TOD has also contributed to the displacement of local residents, the loss of affordable housing in newly developed areas close to transit stations and stops, and the disruption of local business operations. It has been marred by overly complex development processes, financial and construction challenges, and unrealistic expectations for civic renewal. In this chapter, we discuss the characteristics of TOD, the current state of practice in cities around the world, and barriers to its implementation. At the end of the chapter, we present more detail on our two-year study on TOD implementation and outline the remaining chapters of this volume.

Keywords High density • Transportation • Mixed use • Neighborhoods • Walkable

Transit-oriented development has quickly become a popular concept among planners, developers, and engineers because it combines mass transit technology, efficient transportation, and high-density development. Politicians and "city boosters" have adopted TOD as a redevelopment

R. Thomas, L. Bertolini, *Transit-Oriented Development*,
https://doi.org/10.1007/978-3-030-48470-5_1

strategy, while some communities and activists have fought it because of its effects on the displacement of local residents, the loss of affordable housing in newly developed areas close to transit stations and stops, and the disruption of local business operations.

Like most planning concepts, TOD is neither a savior nor a villain. It is a complex process and strategy that can contribute to more sustainable transportation patterns, decreased emissions, and enhanced regional connectivity, but has also been marred by overly complex development processes, financial and construction challenges, and unrealistic expectations for civic renewal. This book explores TOD as a policy concept and is based in part on an in-depth study conducted at the University of Amsterdam (2012–2014). The goal was to study how TOD has been implemented in a range of international case studies with the intent of spurring a breakthrough in TOD in the Dutch context.

In this chapter, we discuss the characteristics of TOD, the current state of practice in cities around the world, and barriers to its implementation. At the end of the chapter, we present more detail on the study and outline the remaining chapters of this volume.

What Is TOD and Why Do Cities Use It?

While most readers of this volume may have some understanding of transit-oriented development and would like to learn more, we would like to ensure everyone has the same understanding of the term.

TOD includes high-density, mixed-use developments located within close proximity to public transportation infrastructure. The term "transit-oriented development" was coined by Peter Calthorpe and detailed in his seminal volume *The Next American Metropolis: Ecology, Community, and the American Dream* (1993). The US-based Center for Transit-Oriented Development, a national clearinghouse on the topic, is the result of a collaboration with the Center for Neighbourhood Technology and Strategic Economics and is funded by the US government. The CTOD definition of the concept is:

> *a type of community development that includes a mixture of housing, office, retail and/or other commercial development and amenities integrated into a walkable neighborhood and located within a half-mile of quality public transportation.* (CTOD 2019)

If this sounds familiar, it is because many cities around the world used to be built this way before the advent of the car. Developing pedestrian-oriented, small-scale urban neighborhoods and streets was commonplace until the 1950s, and these places can be experienced in many historic neighborhoods around the world. For this reason, Calthorpe considers himself "a reviver rather than an originator of ideas" (Newman 1991). As a reaction to decades of car-oriented development, the modern vision of TOD began as a rail-based concept with development focused around station areas, with the basic premise that re-orienting urban development toward more dense corridors had the potential to preserve land, decrease car use, curb urban sprawl, and accommodate regional growth.

CTOD (2007) suggests that demographic shifts, and accompanying shifts in housing preferences, are another reason that cities might invest in TOD: the fastest-growing household type is no longer the two-parent, two-child household, which is now just 25 percent of the total number of households in the US and in Canada. More rapid growth in households made up of couples without children, single parents, people living alone, seniors, and immigrants has resulted in a vastly increased demand for public transit; historically, all of these household types have taken public transit at higher rates than the two-parent, two-child household. A national study showed that by 2030, 40 percent of households in the US were looking for high-density housing near transit (CTOD 2007). People living in areas close to TOD have lower car ownership and smaller household sizes (Ewing and Cervero 2010).

TOD has now evolved to include small-scale developments such as walking paths, cycling parking and trails, and public spaces supporting not just high-capacity metropolitan railway use but also local buses, streetcars, and non-motorized travel modes such as walking and cycling (TransLink 2012). TransLink, the transportation authority in Vancouver, Canada, has broadened the term to "transit-oriented communities" to signify that higher densities, mixed use, high-quality urban design and other characteristics need not be attached to large-scale station area developments, but can in fact be integrated at a variety of scales across a region. TOD "has evolved into a regional or 'network' approach in Europe, relying on the regional services rather than Light Rail" (Conesa 2018, 120). In some cities and countries, TOD is part of a regional approach to growth management and transportation demand management, which aim to concentrate growth in areas with existing infrastructure, rather than sprawl outward.

Why Do Cities Want to Adopt TOD?

As a concept, TOD has characteristics that appeal to environmentalists, health authorities, economic development agencies, real estate agents, and regional transportation authorities. Schuetz et al. write that "City and county governments (and many voters) see rail transit as the key to reducing congestion, restructuring urban form, creating a livable city, and attracting economic development" (2018, 1673). Indeed, some of the benefits of TOD (CTOD 2019; Cervero 2008) include:

- **Decreased** driving, parking needs for cars, regional congestion, air pollution, and greenhouse gas emissions
- **Increased** transit ridership, transit fare revenue, competition of transit with the car, mobility choices, property values near TOD, and access to jobs
- Walkable communities and compact urban development at densities supporting public transit, walking, and cycling

As low-density, car-oriented regions and countries have adopted TOD principles, they have faced many challenges; case studies on existing TODs often examine their mixed results in achieving the expected benefits. Even TOD in more densely developed countries (e.g., the Netherlands, Hong Kong) have faced challenges in implementation. Successful TOD is widely considered to be dependent upon the five Ds: density, diversity, design, distance to transit, and destination accessibility (Ewing and Cervero 2010). Each of these characteristics influences transit ridership in areas close to TOD.

Density is measured as the variable of interest per unit of area, e.g., population, dwelling units, or employment (Ewing and Cervero 2010). TOD has a lot of potential to concentrate new population and employment growth, but this is more complicated than it sounds, particularly when transit infrastructure is located within a car-oriented, low-density region or one with a limited regional transit network. For example, in Dovey et al.'s (2017) study of six tram corridors in Melbourne, Australia, the authors suggested that if all the sites along the corridors were intensified, up to 45,000 dwelling units could be added. However, only 2500 units have been added so far, as Melbourne continues to sprawl outward. Renne (2017) points out that the US invested billions to build over 4400 passenger rail stations across 39 metropolitan regions by 2010, but only 36 percent achieve a minimum density of 8 units per acre, the minimum requirement to support transit use. He writes that "If every station area achieved a minimum density of 8 units per acre, additional housing could be created for 9 million new Americans" (Renne 2017, 474).

Diversity refers to the number of different land uses in a given area and the degree to which they are represented in land area, floor area, or employment, e.g., using low values for single-use environments. Jobs-to-housing or jobs-to-employment are less frequently used (Ewing and Cervero 2010). Diversity can be difficult to achieve in metropolitan regions that have favored single-use zoning and car-oriented development for decades. For example, in their study of five station areas in Los Angeles, Schuetz et al. note that the community plan in the Highland Park neighborhood "frames development as a challenge to be controlled, rather than a desired outcome" (2018, 1682). The plan stresses the need to separate commercial and residential land uses and calls for increased parking, and a separate TOD neighborhood plan stresses the need to maintain the historic neighborhood character.

Design refers to street network characteristics, e.g., block size, proportion of four-way intersections, and number of four-way intersections per square mile. Occasionally sidewalk coverage, average building setbacks, average street widths, or other characteristics are used. High-quality urban design has become a hallmark of new developments in the US, which often tend to have a "1950s main street" look and feel, including a range of physical characteristics that encourage walking and cycling. Dovey et al. (2017, 262) write that:

> *The values of intensified small-grain mixed-use development have become an integral part of recent urban design movements including New Urbanism, transit-oriented development and main street revitalization programs.*

Ewing and Cervero (2010) found that walking was strongly influenced by characteristics such as intersection density and the number of destinations within walking distance. These characteristics also resulted in a lower number of Vehicle Miles Travelled.

Distance to transit The US metric of the half-mile (800 m) is widely considered to be the maximum distance that people will walk to access high-order transit such as metro/subway trains and Light Rapid Transit (LRT). Most people will walk a quarter-mile (400 m) for local stop service. However, walking preferences vary by destination, trip purpose, gender, age, land use, safety, weather, the price and availability of parking, and the density of jobs and services. There are also differences depending on the geographic and cultural context. Guerra and Cervero (2013) found a quarter-mile to be a more reliable measure of transit use in American cities, while Ker and Ginn (2003) found residents in Perth, Australia, were willing to walk over 0.6 miles (1 km) to a transit station. Some transit users bike to transit stations, lengthening the distance they are willing to travel. A Korean study (Lee et al. 2016) found that in Seoul and Daejeon, transit users would bike just over 1.2 miles (2 km) to stations, while Dutch researchers found that cyclists in the Netherlands would bike up to 2.2 miles (3.6 km) to stations (Le Paix Puello and Geurs 2015).

Destination accessibility is defined as the ease of access to trip attractions, either regionally or locally. Some typical measures include the distance to the central business district and the number of jobs accessible within a certain travel time. For example, Deboosere et al. (2018) found that residents in neighborhoods with high access to jobs and low access to competing workers have the shortest commute times in the Greater Toronto and Hamilton Area. High accessibility among the labor force was associated with increases in population density and job density in the region. The authors' vision of "accessibility-oriented development" balances accessibility between employment opportunities and workers; to foster this balance, planners could zone part of a proposed TOD as residential to provide potential housing for employees or provide new and improved transit options to a variety of neighborhoods. In a study of transit accessibility in socially disadvantaged neighborhoods in the Greater Toronto and Hamilton Area, El-Geneidy et al. (2016) found that residents have equal or better accessibility to jobs in the region, and they have shorter commute times than residents of better-off areas.

Ewing and Cervero (2010) found that the distance to downtown resulted in a lower number of vehicle miles travelled (VMT). In 17 TODs in four metro areas in the US, they found that housing in these developments generated up to a 50 percent reduction in trips by car. In a study in Nord Pas-de-Calais, France, Conesa (2018) found that buses, biking, and walking could allow policy makers to improve accessibility to urban opportunities and smaller towns more affordably than rail-based transit modifications.

To summarize, TODs that possess the five Ds are often more successful, but these can be difficult to achieve given the range of spatial, geographic, and cultural contexts in which they are attempted. Cities and regions around the world have implemented a variety of strategies to support TOD.

What Are Cities and Regions Doing to Implement TOD?

The main step cities and regions have taken to implement TOD has been building more extensive transportation networks in order to support development near transit. Cities such as Los Angeles, Seoul, and Shenzhen have all built extensive rail-based infrastructure in recent years, while other cities are currently integrating bus rapid transit or light rail lines (e.g., Toronto, Ottawa, Portland). Frequent transit networks incorporating local bus routes have become more prevalent in major cities: Kramer (2018) used 17 American and Canadian cities with 15-minute, all-day service in her comparison. Ewing and Cervero wrote that "The general consensus is that transit service headways of 10 minutes are ideal to support a transit lifestyle" (2010, 3). In Denver, the city delineated five types of TOD (downtown, urban center, general urban, urban, suburban) which allows context-specific policies to be put in place along transit lines, such as higher zoning allowances in downtown station areas and broader building setbacks in suburban station areas. After construction of the streetcar route in Portland, Oregon, properties closest to the streetcar were developed at 90 percent of the permitted density compared to 43 percent just a few blocks away (CTOD 2007). Seattle's Central Link LRT, Washington D.C.'s Silver Line, and Charlotte, North Carolina's Blue Line were all planned to guide growth and development along entire corridors (CTOD 2010).

Regional and local plans and policies have also been implemented in many countries. In Europe, anti-sprawl, transit ridership, regional growth, and polycentric growth policies are often combined in a regional approach to shifting travel demand toward transit (Conesa 2018). In Australia, Perth and Melbourne have regional plans and specific TOD policies that emphasize growth along transit corridors and in activity centers (Dovey et al. 2017; Griffiths and Curtis 2017). In the state of New Jersey, the Transit Village Initiative offers incentives, technical assistance, and coordinated agency assistance to communities (Noland et al. 2017). The City of Vancouver has linked new locations for affordable housing to their frequent transit network in their housing strategy (City of Vancouver 2018a, b).

Municipal governments have implemented a range of tools to encourage or allow TOD, such as the transfer of air or development rights, density bonuses, and faster processing of development proposals. These can have some interesting effects depending on the specific context. For example, although there is an assumption that new rail stations will be built in existing high-density neighborhoods, where there is already a good level of transit ridership, this is not the case in China. Yang et al. (2016) demonstrated that in Shenzhen, the municipal government actually chose to build rail lines in relatively undeveloped areas, sometimes even bypassing town centers. In American cities, property values are higher in dense, developed areas so there is more of an incentive for municipalities to encourage development on these sites to generate higher property taxes and for developers to build densely on the sites to generate higher rents in the completed project. But in China, there is no property tax: instead, the government owns the land and leases it long-term to private-sector developers, using the transfer of land use rights to raise revenue. The city then provides density bonuses in two bands (within 200 m and 500 m of the stations), with the highest densities within 200 m of the station to encourage TOD. In Hong Kong, station areas on greenfield sites experienced more rapid population growth after new rail stations opened than those on infill sites, but the average population density of the greenfield neighborhoods was 63.2 percent lower than infill areas (Loo et al.2017).

Many cities now have urban design guidelines to promote walkable neighborhoods around transit. Developers strongly associate transit access with walkability (Guthrie and Fan 2016) and see a lot of market demand for walkable communities; developers, planners, and residents agree on the value of transit and TOD in increasing walkability. Municipal officials see TOD as a way to offer an urban quality of life and a stronger sense of community (Noland et al. 2017). Residents of the four New Jersey TOD neighborhoods in their study noted that they had more opportunities to meet neighbors and local shopkeepers because they walked so much more than they had before moving to the area. However:

> *Participants felt that while the development associated with TOD often takes the form of restaurants and bars, they would prefer to see more local retail establishments selling basic or essential goods.* (Noland et al. 2017, 137)

The authors also noted that residents raised some concerns about pedestrian safety in their neighborhoods, mainly absent or poorly maintained sidewalks and unsafe street crossings. Displacement of local businesses and the location of many TODs in areas that are not necessarily pedestrian or transit-friendly may also be challenges (Chapple and Loukaitou-Sideris 2019).

Despite efforts to support TOD through plans, policies, and guidelines, its implementation and the realization of its many promised benefits are still fraught with difficulties.

Barriers to TOD

Many cities and regions have not seen very promising results from TOD implementation. For example, Yang and Pojani (2017) found a very moderate trend toward increased jobs in station areas in Brisbane, Australia, while population and housing densities were actually lower than in areas without TOD. In the Netherlands, a prolonged economic crisis, the absence of an academic and political consensus on TOD, weak regional governance institutions, an office space surplus in station areas, and a mismatch between the supply and demand of areas for residential development were key barriers to TOD (Lenferink and Van der Stoep 2013; Pojani and Stead 2014; Tan et al. 2014; Thomas and Bertolini 2014). In this section, we present some of the main barriers to TOD implementation.

Zoning and Other Regulatory Issues

Cities that have been planned around the private automobile and suburb-to-city commute have extensive regulations on land use and development favoring the car:

> *TOD (e.g. mixed uses, high densities, reduced parking) is still illegal around station areas in many cities and transit districts, creating a barrier for development.* (Ewing and Cervero 2010, 4)

A good example of this is parking standards. Although development near public transit will presumably decrease automobile use, municipal plans often require a certain number of parking spots per housing unit, and these standards are based on the assumption that every household owns one or more cars. Parking spots in built-up areas are quite expensive to build, and high parking ratios can result in less profitable developments, higher fees, and exactions for developers. Ewing and Cervero (2008) estimated that lowering residential parking ratios by 50 percent for TODs in rail station areas could result in densities that were 20–33 percent higher (depending on the building type). They wrote that "TOD developments that do get built are certainly less affordable and less sustainable than they might be because they are subject to incorrect assumptions about generated traffic impact" (2008, 54).

Chatman (2015) found that the proliferation of on- and off-street parking impacted car ownership around rail stations in New Jersey. Scarce off-street parking resulted in a 40 percent decrease in commuting to work by private automobile; rail only reduced car ownership when combined with low parking availability, and even then residents still used their cars at the same amount for commuting or grocery trips. In the provocatively titled "Does transit-oriented development need the transit?," Chatman suggested that building smaller units with less parking in locations with better bus service and higher employment density might be a better strategy than building TOD to reduce greenhouse gas emissions. Developers found that relaxed parking requirements and higher permitted densities were important in making affordable housing projects viable in TOD areas (Guthrie and Fan 2016).

The size and shape of lots can also be a barrier to TOD, both at the large and small scales. Dovey et al. (2017) found that along Melbourne's tram corridors, small lots had a lower yield and lower profits, and very narrow sites were also resistant to intensification. Most intensification occurred on medium-sized lots, though broad and corner sites were also attractive for intensification and yielded substantially higher densities.

Schuetz et al. (2018) found a number of perplexing conditions in the five Los Angeles station areas they studied; e.g., in some cases the allowable densities for new buildings were lower than existing structures. Zoning incompatibility across all five stations presented real challenges for developers, including complex rules and procedures, uncertainty, and the length of the process. As a result, redevelopment near station areas in Los Angeles takes many years and relies upon a combination of factors:

> *...the probability, form, and timing of transit-oriented redevelopment depends on compatible zoning, strong real estate markets, active engagement by local*

> *government agencies and (in most cases) political support from neighborhood residents.* (Schuetz et al. 2018, 1685)

Policy Consistency and Planning Coordination

Lack of coordination between transit investments and land use at the regional scale (e.g., Deboosere et al. 2018) presents a significant barrier to TOD. Transportation, land use, and housing development and planning are often conducted separately. Different municipal departments or levels of governments, as well as private-sector partners, often have conflicting values and goals (Duffhues and Bertolini 2016). In many cities, the success of TOD is therefore limited by a lack of broader, regional-level approaches to decreasing car use. For example, in Perth, Australia, it has been difficult to increase transit ridership because of ample parking at the rail stations and the lack of transit access to services and amenities in the region outside of TOD areas like the Subiaco Station precinct. Griffiths and Curtis (2017, 406) note that "a single TOD precinct cannot function fully as TOD without such metropolitan wide changes." Ewing and Cervero (2010) found that residents of TOD who used transit for their commute to work did so at two to five times the rate of other commuters in the region. But transit use ranged from 5 to 50 percent depending on relative travel times by car and the extent of transit service in the region:

> *Systems that generate the highest commute ridership also have the highest percentage of regional jobs accessible by fast transit.* (Ewing and Cervero 2010, 3)

Lack of agreement on the sites to be developed through TOD (e.g., Thomas et al. 2018) has also been a significant barrier to development of station areas. A municipality or transportation authority may have to expropriate land in station areas and/or consolidate it for large-scale developments, which is difficult and time-consuming.

If TOD is to achieve regional objectives such as decreased car use or urban sprawl, it needs to be applied beyond isolated station areas and in coordination with tools, policies, and plans regulating urban growth, development of employment areas, and housing location.

Cost

The infrastructure for rail-based transportation is quite expensive and time-consuming to construct in built-up areas. Mid-sized cities and regions usually cannot afford rail infrastructure, which is one reason they tend to support bus rapid transit and LRT instead. Even in large metropolitan areas or countries with larger budgets devoted to transportation infrastructure, governments often partner with the private sector to fund station development.

The complexity of TODs also adds to the cost for developers. Interviews with residential and commercial developers in Minneapolis-St. Paul revealed that increased costs, complexity, and length of regulatory processes in TOD areas made it more difficult for developers to build in these areas (Guthrie and Fan 2016). Financing for affordable housing in TODs was difficult to obtain, but nevertheless the authors found "pent-up demand for transit access" among developers (ibid., 108). Overall, they found an "artificially constrained supply of TOD projects" (ibid., 111) due to the constraints mentioned above. Chapple and Loukaitou-Sideris (2019) noted that in many American TODs, high land costs for purchasing and assembling land, lack of developable land near stations, and weak real estate markets are persistent barriers to mixed-use development following the construction of transit stations.

To make building in station areas more attractive to developers, a number of financial tools have been used, such as tax increment financing (TIF), which allows municipal governments to use the expected increase in tax revenues to fund development. Density bonuses have allowed developers to build higher in station areas and transit corridors, generating more efficient and profitable projects.

Public Opposition

During the railway era (1850–1920), countries like the US, Canada, and the UK were largely rural, with very little land devoted to urban areas. In fact, a key motivation behind the Victorian-era cross-country rail projects in both the US and Canada was to connect these dispersed areas so that natural resources and manufactured goods could be quickly and easily transported. Municipalities did not have planning regulations to direct where railway lines could or could not go, and the public generally supported rail projects for providing national growth or economic benefits.

While cross-country railway lines were supported by national governments, streetcar lines within cities were developed and operated by small private companies. In larger cities, metro lines and property development were integrated, sometimes within one company, as was the case with the London underground (Levinson 2008) and is still the case in Asian cities like Tokyo (Chorus and Bertolini 2016) or Hong Kong (Loo et al. 2017). The linear land development pattern that supported streetcar or subway routes back then is the basis for neotraditional development today, including TOD. Streetcar routes were developed by private companies who sold the lots along their proposed rail line in order to fund its construction (Renne 2017).

But during the freeway era (Muller 2004) beginning in 1945, large-scale infrastructure projects (particularly highways and public housing projects) resulted in massive displacement and destruction of low-income communities. These residents and others began to question the benefits of large-scale infrastructure projects, and in many cases, community activists were successful in stopping them altogether: activists in North American cities like Vancouver, San Francisco, Toronto, and New York and in European cities like Amsterdam, Copenhagen, Zurich, and Munich (Bratzel 1999; Switzer et al. 2015) halted freeway construction. Land use planning became firmly established in municipalities, and most municipalities and regions had comprehensive plans by the 1960s. Federal and state/provincial transportation authorities, who were usually the funders behind large-scale transportation infrastructure projects, had to respect local regulations on where and what could be built within cities (DiMento and Ellis 2012). It is now commonplace for neighborhood residents to oppose transportation infrastructure, whether it is bike lanes, bus lanes, LRT lines, or rail stations. Residents may also oppose changes such as increased density, urban design characteristics, or decreased parking levels, which are commonly incorporated in TOD projects (Filion and McSpurren 2007). Though public support may not be sought or required in some countries, opposition to projects is often noted (e.g., Yang et al. 2016).

Loss of Affordable Housing

The loss of affordable housing in TOD projects is a significant barrier to more extensive implementation, although the national context is important here. For example, in European and Asian cities with a high percentage of public housing stock, this issue is not as relevant. Indeed, in some of these cities, TOD might be an integral part of social housing development, as has been the case with the development of rail-based new towns in Stockholm and Hong Kong (Cervero 1998; Loo et al. 2017). But in the US, Canada, Australia, and the UK, where the private market supplies most housing, rising rents in transit corridors and station areas is a significant area of concern. In the US, developers perceive TOD to be a "niche market," meaning that it appeals to a small but distinct percentage of the population (Guthrie and Fan 2016), such as younger households who tend to have higher educational levels and incomes. The increasing appeals of car-free lifestyles and walkable communities, combined with their relatively scarcity, have combined to price out low-income households in TODs and transit corridors (e.g., along LRT lines).

In 2007, CTOD wrote that American cities needed to act quickly to preserve affordable housing near transit because low-income households were more likely to take transit and they had low car ownership. Kramer (2018) conducted an analysis of 17 large American and Canadian metro regions to determine the relationship between their frequent transit networks (defined as providing 15-minute frequency between the hours of 7 am to 7 pm) and socioeconomic status. She found that areas within 500 m of a local transit route or within 1 km of a rapid transit station have "older, denser housing stock, more rental, and are home to more ethnically and racially diverse populations, with lower incomes" but that housing does not consistently cost less in these areas. In many cases, households must choose between affordable transportation and affordable housing. Locally owned shops are also at risk: Chapple et al. (2017) found commercial gentrification in Los Angeles and San Francisco, as rising rents put increased pressure on local merchants. The interviews revealed that rising rents, rather than transit access, were the main reason their neighborhoods were changing.

In their Mixed-Income Housing Action Guide, CTOD (2009) suggested that community advocates and policy and decision makers needed to collect data on the demographics, housing stock, market conditions, land supply, and policy environment in order to understand how to preserve affordable housing in areas where TOD is being considered. They presented a range of policy tools that could be used in both "warm" and "cold" real estate markets, including reduced parking requirements, tax forgiveness for affordable housing, and fee waivers, reductions, and deferrals. For example, in Cleveland, Ohio, the Cleveland Regional Transit Authority partnered with local development corporations to promote development that included market-rate and affordable housing options along the Health Line BRT (CTOD 2010). In San Francisco, a partnership of five non-profit organizations and two regional foundations created the Bay Area Transit-Oriented Affordable Housing Fund to buy properties near transit for affordable housing (CTOD 2011). More examples of equitable TOD (eTOD) will be discussed in Chap. 4.

Conclusion

Transit-oriented development has quickly become a popular concept among planners, developers, and engineers. However, like most planning concepts, TOD has its challenges. Although transit-oriented communities are certainly not new, returning to this style of development poses challenges, particularly in countries where most of the land was developed and settled during the automobile era.

Once thought of as a rail station-based concept, TOD has now evolved to include small-scale developments such as walking paths, cycling parking and trails, and public spaces supporting not just high-capacity metropolitan railway use but also local buses, streetcars, and non-motorized travel modes such as walking and cycling. TODs that have the five Ds (density, diversity, design, distance to transit, destination accessibility) are often more successful. Cities and regions around the world have implemented a variety of strategies to support TOD, including more extensive transportation networks, regional and local plans and policies, tools to encourage or allow TOD, and urban design guidelines to promote walkable neighborhoods around transit.

However, TOD implementation, and the realization of the many promised benefits, is still challenging. Despite these efforts at implementing TOD, many cities and regions have not seen very promising results from its implementation. Cities that have been planned around the private automobile and suburb-to-city commute have extensive regulations on land use and development favoring the car, e.g., parking requirements, and land use zoning stipulating the separation of residential and commercial land uses. Lack of coordination between transit investments and land use at the regional scale presents a significant barrier to TOD. The infrastructure for rail-based transportation is quite expensive and time-consuming to construct in built-up areas. Public opposition to TOD projects is commonplace. The loss of affordable housing in TOD projects is a significant challenge demanding a solution if transit-accessible neighborhoods are to remain inclusive and diverse.

The Structure of This Volume

In the following chapters, we will delve into an in-depth study of TOD. Our study, based in the Netherlands, examined successful TOD implementation with the intent of bringing these ideas back to the Dutch context. While our study focused on bringing ideas from other countries to the Netherlands, we feel that the research findings are applicable in other contexts. Our case studies were international city-regions and were chosen to represent a range of "successes" and "failures" in TOD implementation; they are, in general, cases of interest to transportation and land use planners. As discussed in the previous section, many cities and countries face similar barriers in implementing TOD, so the lessons from other contexts could be useful. The methods that we used can be easily used in other contexts, particularly since in-depth case studies on TOD are common: a structured comparison of completed case studies would be possible in many national contexts, as would policy transfer from international contexts to the local municipal or regional government.

Funded by the Netherlands Organisation for Scientific Research (Nederlandse Organisatie voor Wetenschappelijk Onderzoek), our two-year study (2012–2014) sought to understand how to force a breakthrough in the implementation of TOD in the Netherlands. The study had three parts, which were conducted simultaneously by three research teams. Part 1 examined actors, policies, and governance arrangements that have contributed to TOD implementation in international cases and was conducted by researchers at the University of Amsterdam. Part 2 identified and tested public and/or private financial arrangements and tools to encourage TOD implementation and took place at Radboud University in Nijmegen. Part 3 explored the design characteristics of successful TOD projects and was conducted by researchers at the Technical University of Delft. This volume focuses on Part 1 of the study, which has been published in four peer-reviewed articles:

- Thomas, R. and Bertolini, L. 2014. Beyond the case study dilemma in urban planning: Using a meta-matrix to distil critical success factors in transit-oriented development. *Urban Policy and Research*,

32(2): 219–237. doi: https://doi.org/10.1080/08111146.2014.882256

- Thomas, R. and Bertolini, L. 2015. Policy transfer among planners in transit-oriented development. *Town Planning Review*, 86(5): 537–560. DOI: https://doi.org/10.3828/tpr.2015.32
- Thomas, R. and Bertolini, L. 2015. Defining critical success factors in TOD implementation using rough set analysis. *Journal of Transport and Land Use,* 10(1): 139–154. DOI: https://doi.org/10.5198/jtlu.2015.513
- Thomas, R., Pojani, D., Lenferink, S., Bertolini, L., Stead, D. and van der Krabben, E. 2018. Is transit-oriented development (TOD) an internationally transferable policy concept? *Regional Studies*. DOI: https://doi.org/10.1080/00343404.2018.1428740

In this book, we have reworked these findings and expanded on them by integrating new studies on TOD and persistent challenges to implementation that are being addressed through approaches focusing on equitable development.

In Chap. 2, we discuss the theoretical approach to learning from other cities/countries, including some of the advantages and disadvantages to transferring policy ideas from one country to another. We focus on learning from case studies, including both successes and failures, and ways in which planners and policy makers have learned from others.

In Chap. 3, we present 11 TOD case studies and then compare the cases using meta-analysis. We discuss how we chose the case studies, present profiles of each city-region, and present results from a meta-analysis of the case studies, including critical success factors in the implementation of TOD. We discuss how we used these research findings in policy learning workshops for planners and policy makers in two Dutch city-regions.

Chapter 4 reflects on some of the ongoing problems with TOD and their solutions, with examples from municipalities other than the 11 case studies profiled in Chap. 3 and a particular emphasis on equitable TOD. Gentrification in station areas and corridors, tools and strategies to enable TOD, building a collaborative practice, and generating support from higher levels of government are discussed.

In Chap. 5, we summarize the practice of TOD, issues with its implementation, and lessons for planners, developers, and communities who want to build more equitable, collaborative TODs that help reorient their regions toward sustainable development.

References

Bratzel, S. (1999). Conditions of success in sustainable urban transport policy: Policy change in 'relatively successful' European cities. *Transport Reviews, 19*(2), 177–190. https://doi.org/10.1080/014416499295600.

Center for Transit-Oriented Development (CTOD). (2007). *Why transit-oriented development and why now?* USA: Reconnecting America.

Center for Transit-Oriented Development (CTOD). (2009). *The mixed-income housing TOD action guide.* USA: Reconnecting America.

Center for Transit-Oriented Development (CTOD). (2010). *Transit corridors and TOD: Connecting the dots.* USA: Reconnecting America.

Center for Transit-Oriented Development (CTOD). (2011). *Planning for TOD at the regional scale: The big picture.* USA: Reconnecting America.

Center for Transit-Oriented Development (CTOD). (2019). *Frequently asked questions.* http://ctod.org/faqs.php. Accessed 16 Feb 2019.

Cervero, R. (1998). *The transit Metropolis: A global inquiry.* Washington, DC: Island Press.

Cervero, R. (2008). *Effects of TOD on housing, parking and travel.* Centre for Transit-Oriented Development: Urban Land Institute, Transit Cooperative Research Program Report 128.

Chapple, K., & Loukaitou-Sideris, A. (2019). *Transit-oriented displacement or community dividends?* Cambridge, MA/London, UK: MIT Press.

Chapple, K., Loukaitou-Sideris, A., Gonzalez, S. R., Kadin, D., & Poirier, J. (2017). *Transit-oriented development and commercial gentrification: Exploring the linkages.* UC Connect: University of California Berkeley Center for Community Innovation and University of California Los Angeles Lewis Center.

Chatman, D. (2015). Does transit-oriented development need the transit? *Access, 1*(47), 16–21.

City of Vancouver. (2018a). *Housing Vancouver strategy.* Vancouver, BC: City of Vancouver.

City of Vancouver. (2018b). *Affordable housing choices interim zoning policy.* Vancouver, BC: City of Vancouver.

Chorus, P., & Bertolini, L. (2016). Developing transit-oriented corridors, insights from Tokyo. *International Journal of Sustainable Transportation, 10*(2), 86–95.

Conesa, A. (2018). The accessibility assessment and the regional range of transit-oriented development: An application of schedule accessibility measures in the Nord Pas-de-Calais region. *The Journal of Transport and Land Use, 11*(1), 119–141.

Deboosere, R., El-Geneidy, A., & Levinson, D. (2018). Accessibility-oriented development. *Journal of Transport Geography, 70*, 11–20.

DiMento, J., & Ellis, C. (2012). *Changing lanes: Visions and histories of urban freeways.* Cambridge, MA: MIT Press.

Dovey, K., Pike, L., & Woodcock, I. (2017). Incremental urban intensification: Transit-oriented re-development of small-lot corridors. *Urban Policy and Research, 35*(3), 261–274. https://doi.org/10.1080/08111146.2016.1252324.

Duffhues, J., & Bertolini, L. (2016). From integrated aims to fragmented outcomes. Urban intensification and transportation planning in the Netherlands. *Journal of Transport and Land Use, 9*(3), 15–34.

El-Geneidy, A., Buliung, R., Diab, E., van Lierop, D., Langlois, M., & Legrain, A. (2016). Non-stop equity: Assessing daily intersections between transit accessibility and social disparity across the greater Toronto and Hamilton area (GTHA). *Environment and Planning B: Planning and Design, 43*(3), 540–560.

Ewing, R., & Cervero, R. (2008). Travel and the built environment. *Journal of the American Planning Association, 76*(3), 265–294.

Ewing, R., & Cervero, R. (2010). *Effects of TOD on housing, parking, and travel.* Transportation Research Board: Transit Cooperative Research Program Report 128.

Filion, P., & McSpurren, K. (2007). Smart growth and development reality: The difficult co-ordination of land use and transport objectives. *Urban Studies, 44*(3), 501–523.

Griffiths, B., & Curtis, C. (2017). Effectiveness of transit oriented development in reducing car use: Case study of Subiaco, West Australia. *Urban Policy and Research, 35*(4), 391–408. https://doi.org/10.1080/08111146.2017.1311855.

Guerra, E., & Cervero, R. (2013). Is a half-mile circle the right standard for TODs? *Access, 42*, 18–21.

Guthrie, A., & Fan, Y. (2016). Developers' perspectives on transit-oriented development. *Transport Policy, 51*, 103–114.

Ker, I., & Ginn, S. (2003). Myths and realities in walkable catchments: The case of walking and transit. *Road and Transport Research, 12*(2), 69–80.

Kramer, A. (2018). The unaffordable city: Housing and transit in North American cities. *Cities, 83*, 1–10. https://doi.org/10.1016/j.cities.2018.05.013.

Le Paix Puello, L., & Geurs, K. (2015). Modelling observed and unobserved factors in cycling to railway stations: Application to transit-oriented-developments in the Netherlands. *European Journal of Transport Infrastructure Research, 15*, 27–30.

Lee, J., Choi, K., & Leem, Y. (2016). Bicycle-based transit-oriented development as an alternative to overcome the criticisms of the conventional transit-oriented development. *International Journal of Sustainable Transportation, 10*(10), 975–984.

Lenferink, S., & Van der Stoep, H. (2013, July 15–19). *Innovative governance and finance strategies for implementing Dutch transit-oriented development.* Paper presented at the 5th AESOP/ACSP Joint Congress, Dublin, Ireland.

Levinson, D. (2008). Density and dispersion: The co-development of land use and rail in London. *Journal of Economic Geography, 8*(1), 55–77.

Loo, B. P. Y., Cheng, A. H. T., & Nichols, S. (2017). Transit-oriented development on greenfield versus infill sites: Some lessons from Hong Kong. *Landscape and Urban Planning, 167*, 37–48.

Muller, P. O. (2004). Transportation and urban form: Stages in the spatial evolution of the American metropolis. In S. Hanson & G. Giuiano (Eds.), *The geography of urban transportation* (3rd ed., pp. 59–85). New York: Guildford Press.

Newman, M. (1991). Focus: Sacramento, Calif.; A Transit-Oriented Approach to Suburbia. *The New York Times*, November 10, 1991.

Noland, R. B., Weiner, M. D., DiPetrillo, S., & Kay, A. I. (2017). Attitudes towards transit-oriented development: Resident experiences and professional perspectives. *Journal of Transport Geography, 60*, 130–140.

Pojani, D., & Stead, D. (2014). Ideas, interests, and institutions: Explaining Dutch transit oriented development challenges. *Environment and Planning A, 46*, 2401–2418. https://doi.org/10.1068/a130169p.

Renne, J. L. (2017). Make rail (and transit-oriented development) great again. *Housing Policy Debate, 27*(3), 472–475.

Schuetz, J., Giuliano, G., & Shin, E. J. (2018). Does zoning help or hinder transit-oriented development? *Urban Studies, 55*(8), 1672–1689.

Switzer, A., Bertolini, L., & Grin, J. (2015). Understanding transitions in the regional transport and land-use system: Munich 1945–2013. *Town Planning Review, 86*(6), 699–723.

Tan, W. G. Z., Janssen-Jansen, L. B., & Bertolini, L. (2014). The role of incentives in implementing successful transit-oriented development strategies. *Urban Policy and Research, 32*, 33–51. https://doi.org/10.1080/08111146.2013.832668.

Thomas, R., & Bertolini, L. (2014). Beyond the case study dilemma in urban planning: Using a meta-matrix to distil critical success factors in transit-oriented development. *Urban Policy and Research, 32*(2), 219–237. https://doi.org/10.1080/08111146.2014.882256.

Thomas, R., Pojani, D., Lenferink, S., Bertolini, L., Stead, D., & van der Krabben, E. (2018). Is transit-oriented development (TOD) an internationally transferable policy concept? *Regional Studies, 52*(9), 1201–1213. https://doi.org/10.1080/00343404.2018.1428740.

TransLink. (2012). *Transit-oriented communities guidelines. Creating more livable places around transit in Metro Vancouver*. Vancouver: TransLink.

Yang, K., & Pojani, D. (2017). A decade of transit oriented development policies in Brisbane, Australia: Development and land-use impacts. *Urban Policy and Research, 35*(3), 347–362.

Yang, J., Chen, J., Le, X., & Zhan, Q. (2016). Density-oriented versus development-oriented transit investment: Decoding metro station location selection in Shenzhen. *Transport Policy., 51*, 93–102.

CHAPTER 2

Policy Learning: How Planners Learn from Each Other

Abstract Planning is a multidisciplinary profession, drawing upon many different fields for its theory, ideas, frameworks, and even institutions. Planning has been built upon the central tenets of mutual learning, analysis, policy making, and direct action to improve communities, and its practitioners have applied theories from many different fields. Planners use a range of methods to develop community plans, draft policy, and deliver programs and services. The case study approach is one of the most frequently used methodologies in policy learning and development. This chapter outlines the ways in which planners learn from each other in practice and research, with a particular focus on the use of case studies and policy learning processes. Challenges with these approaches are discussed, along with suggestions on how to meaningfully learn from other contexts. The chapter sets the stage for Chap. 3, which presents and analyzes 11 case studies in TOD.

Keywords Case study • Meta-analysis • Policy learning • Policy transfer

Planning is a multidisciplinary profession, and as such it draws upon many different fields for its theory, ideas, frameworks, and even institutions. The earliest professional planning institutions granted membership to those working in allied professions such as engineering and architecture. John

R. Thomas, L. Bertolini, *Transit-Oriented Development*,
https://doi.org/10.1007/978-3-030-48470-5_2

Friedmann, in *Planning in the Public Domain* (1987), presented four major intellectual traditions of the discipline (social reform, policy analysis, social learning, and social mobilization), each of which contributed tools, ideas, and theories into planning. Planning has been built upon the central tenets of mutual learning, analysis, policy making, and direct action to improve communities, and its practitioners have applied theories from many different quarters.

Observing communities, measuring and analyzing their characteristics using data have been essential practices for decades; since the 1970s, life histories, focus groups, community forums, and many other methods have balanced the quantitative data gathered in large datasets, such as national-level censuses and transportation data on travel patterns. While planners use a range of methods to develop community plans, draft policy, and deliver programs and services, the case study approach is one of the most frequently used methodologies in policy learning and development.

This chapter outlines the ways in which planners learn from each other in practice and research, with a particular focus on the use of case studies and policy learning processes. Challenges with these approaches are discussed, along with suggestions on how to improve upon them. The chapter sets the stage for Chap. 3, which presents and analyzes 11 case studies in TOD.

Learning from Case Studies

Planners regularly study policies from other places, adapting them to suit local governance, policy, and cultural contexts. Bent Flyvbjerg proposed that learning from context-specific case studies allowed practicing planners to progress from "rule-based beginners to virtuoso experts" (2006, 221). No matter what type of expertise a person has, Flyvbjerg argued that every expert "operates on the basis of intimate knowledge of several thousands of concrete cases in their areas of expertise" (2006, 222). The planner working at a municipality on a zoning bylaw regularly has undoubtedly worked on a similar regulation before and can also access bylaws in other municipalities to learn from. Planners from municipalities and regional planning organizations also host foreign planners and visit their counterparts in foreign countries to learn about specific policies or programs. For example, the City of Toronto was so strongly influenced by the design of "pedestrian scrambles" in cities such as Tokyo and London that it installed its first such intersection in 2008; the City continues to draw upon foreign examples to develop policy (Goldberg-Miller 2018). Tan (2011, 145) described the exchange of ideas between transportation planners in Perth, Australia, and foreign academics and planners:

> *Professional organizations are known to arrange at least one local and foreign study trip annually to learn from what they consider as good examples of land use and transport developments and interact with local experts.*

As Flyvbjerg's own research on power and decision making in Aalborg, Denmark, illustrates, academic planners conducting research into a long-term plan, a planning process, or a complex phenomenon often turn to the case study. Integrating a combination of analytical methods such as descriptive statistics, interviews, participant observation, GIS, and analyzing built form, case studies often provide rich data from which to theorize and make recommendations for planning interventions or improved processes. As Maxwell (2004, 248) puts it:

> *Process theory…deals with events and the processes that connect them; it is based on an analysis of the causal processes by which some events influence others. Process explanation, since it deals with specific events and processes, is less amenable to statistical approaches. It lends itself to the in-depth study of one or a few cases or a relatively small sample of individual and to textual forms of data that retain the chronological and contextual connections between events.*

The ability of the case study methodology to integrate multiple methods makes it particularly useful in planning, as a multidisciplinary field integrating geography, sociology, public health, psychology, economics, and other disciplines that contribute their own methods and approaches.

Single-case studies can often provide unique opportunities for researchers to focus on the intricacies of specific cities, which tells a compelling story of conflicts and power dynamics. As Ragin (1987, 34) writes about case selection:

> *It is their particularity—the fact that they are instances of significant events or phenomena—that attracts the attention of the investigator. Sometimes, there is only one or two or a small handful of such instances.*

Although the case study methodology is inconsistent with the requirements of statistical sampling procedures which are fundamental to generalization (Yin 1981; Schofield 2002), Yin (1994) and Flyvbjerg (2001) both indicate the value of using atypical cases in analytic generalization: the ability of a theory to be tested in a similar theoretical setting to further define its explanatory power. Schofield (2002) gives an example of generalizing from an atypical case to a typical one. Prominent case study researchers also propose the multiple-case study (e.g., Yin and Heald 1975; Yin 1994) as a way to develop and test theories and comparison of cases for more generalized theory-building. In their comparison of sustainable mobility in two cities and their urban planning and development, Naess et al. (2011, 290) argued that:

> *Copenhagen and Oslo could arguably be characterised as 'critical cases' of urban sustainability, understood to mean that any main shortcomings and barriers to sustainable urban development and sustainable mobility in these cities are also likely to be present among European cities with lower sustainability ambitions and achievements.*

Single-case studies are regularly explored in full-length volumes in order to draw broader conclusions about planning. For example, Planetizen's Top 10 Planning Books of 2018 listed *High-Risers: Cabrini-Green and the Fate of American Public Housing* (Ben Austen) and *The Poisoned City: Flint's Water and the American Urban Tragedy* (Anna Clark); their 2017 list featured *Never Built New York* (Greg Goldin and Sam Lubell); and 2015 listed *How Paris Became Paris: The Invention of the Modern City* (Joan DeJean). Clark's volume will undoubtedly be used to understand the context of municipal decisions complicated by state and federal jurisdiction that had dangerous effects on public health, while Austen's offers a cautionary tale on large-scale urban renewal and the destruction of the urban fabric. Canadian instructors regularly use resources such as *Remember Africville* (1989), a film documenting the displacement of an African Canadian community during the urban renewal era, while British instructors will likely use the modern case of the Grenfell Towers apartment fire to educate future generations.

Single-Case Studies and Policy Learning

Although single-case studies provide a wealth of learning opportunities for planners, there are a number of problems when it comes to translating their findings into recommendations for planning practice. Because their findings cannot be statistically generalizable, single-case studies often do not provide enough actionable recommendations for policy development. In some cases, the political, geographic, or demographic context of a case study makes it difficult to apply its findings in a different city or country. Contextual issues often act as a barrier to implementation of key policies or programs in other jurisdictions. Stakeholders in planning processes often find it so difficult to separate context from policy ideas that they object to "outside" ideas, insisting that a policy or project might work *there*, but never *here*. One solution to these problems with single-case studies is more systematic learning from multiple cases through comparison or meta-analysis.

Case Study Comparison and Cross-Case Techniques

Case study comparison can be used to draw out similarities between cases. Case studies in TOD have been compared in several volumes: Newman (2007) focused on Australian cities and Bertolini and Spit (1998) on European cities; Cervero (1998), Curtis et al. (2009), and Bertolini et al. (2012) used international cases. Newman (2007) and Curtis et al. (2009) attempted to find common elements between the cases, with tables comparing land use and transport policies, processes, funding mechanisms, and leadership across cases. Bertolini and Spit (1998) identified recurring dilemmas, while Cervero (1998) created a typology of TOD cases and identified factors contributing to their success as well as challenges they encountered. For case studies to generate more generalizable research findings, research that synthesizes completed case studies in an attempt to find common concepts, issues, or tools could be the next step. Given the large number of planning case studies that exist, it should be easy for researchers to find and compare a number of similar cases.

Sandelowski et al. (1997) argue that projects that aim to synthesize the findings of case studies should be recognized as an important avenue in knowledge development. Miles and Huberman (1994) present two reasons to do cross-case analysis: to enhance generalizability and to deepen understanding and explanation, particularly defining the specific conditions under which a finding will occur. Khan and VanWynsberghe (2008) proposed cross-case techniques "as a mechanism for mining existing case studies so that knowledge from cases can be put into service for broader purposes," for example, gathering evidence to modify or reinforce existing policy. Andersson et al. (2002) noted that only through systematic comparisons was it possible to say anything definitive about the characteristics and types of cases they studied.

One challenge in synthesizing case studies for comparison is the integration of data as diverse as interviews, surveys, direct observation, and policy analysis. Dixon-Woods et al. (2005) explored techniques to synthesize evidence from primary case study reports, including narrative summary, thematic analysis, qualitative metasynthesis, content analysis, and cross-case techniques. A number of these methods are able to reduce qualitative data significantly, albeit at a loss of the unique richness inherent in this type of data. Both Dixon-Woods et al. (2005) and Sandelowski et al. (1997) express concern that researchers need ways to incorporate qualitative data into systematic reviews; the latter argues that the key to understanding several studies in relation to each other may lie in recognizing how they all reprise familiar stories.

Cross-case techniques can allow researchers to mine case study data for common elements. In Naess et al.'s comparison of Copenhagen and Oslo (2011, 287):

> *The comparison across case cities has focused on common traits as well as differences and has attempted to explain both. These explanations were first made within each case. Explanations of similarities and differences were thus based on a study of generative mechanisms and cross-case comparison of these mechanisms...*

Other transportation researchers have used systematic case study comparison. Van Egmond et al. (2003) compared the public transport systems in 22 European cities, initially using case reports and relevant documents, and data gathered in a systematic way by local experts in each city. Walter and Scholz (2007) assessed collaborative urban transport projects in five cities using reports, relevant documents, and interviews with local experts. Papa and Bertolini (2015) compared six Western European cities to determine the link between TOD and urban form, specifically the links between the rail network and the spatial distribution of jobs and population. They used correlation analysis to understand the connection between the urban structure and accessibility, using GIS to help visualize the results and provide a spatial analysis. In a comparison of shared mobility in 13 metropolitan areas, the US Department of Transportation (2018) systematically assessed activities and planning documents within 13 metropolitan areas and conducted interviews with agencies in several of the metro areas. They found that agencies were uniquely positioned to coordinate shared mobility, but faced several challenges to integrating it into their existing approaches; they were able to make several recommendations for other regions. Ewing and Cervero (2010) used 62 studies of travel and the built environment in a quantitative meta-analysis, calculating elasticities between travel variables and built environment variables. The studies used a variety of statistical methods and capture the effects of more than one "D" variable (density, diversity, design, destination accessibility, and distance to transit). They were able to conclude that destination accessibility (e.g., distance to downtown) was most important in reducing travel by car; street network design (intersection density, street connectivity) was also significant. Andersson et al. (2002) found that cross-case analysis was useful in identifying common themes in case studies, but their ability to generalize results was limited by the fact that the nine cases were very different and reflected observations from extensive, but non-random, data. Choosing similar cases, Nijkamp et al. (2002) took their analysis a step further, first identifying within-case and cross-data patterns.

One cross-case technique, meta-analysis, aims to derive common elements from a series of completed case studies, often in order to identify transferable lessons in the form of conditional statements, which would specify the conditions under which the statements are valid. Nijkamp et al. (2002, 1873) wrote that "Meta-analysis aims to offer a statistical underpinning for the comparison of studies within the same broad research field." A major benefit of meta-analysis is that completed case studies can be used, decreasing the time-consuming fieldwork that is characteristic of case study research. Cases can then be compared, evaluated, and ranked on the basis of well-defined criteria or performance measures, and key factors that are responsible for differing results across similar studies can be identified. Baaijens and Nijkamp (2001, 822) observe that meta-analysis "is particularly suitable in cases where research outcomes are to be judged or compared (or even transferred to other situations), when there are no controlled conditions." This sounds remarkably like a practicing planner exploring techniques, tools, and processes from other jurisdictions in the hopes of introducing an innovative approach in their own city.

Miles and Huberman (1994) describe meta-matrices, which can be used in a meta-analysis to allow the researcher to see all of the qualitative and quantitative case study data together in one place. The process of making the meta-matrix, narrowing down, and categorizing the categories and entries helps the researcher develop explanations for the patterns within and across the cases. Organizing the entries by date, for example, allows the researcher to trace the flow of events in a process in a systematic and transparent way. One limitation is that meta-matrices can only be used in cases that are similar to each other. A case-oriented or variable-oriented analysis can be used (Ragin 1987): case-oriented analyses determine the distinctive configuration and flow of events within each case and then compare across cases, while variable-oriented analyses determine the mutual influence of a specified set of variables. However, the complexity of social processes means that there are often too many factors to isolate (Khan and VanWynsberghe 2008).

Case study comparison, or systematic cross-case techniques such as meta-analysis, can evidently provide planners with more generalizable results than single-case studies. But how do case study findings make their way into policies, plans, and tools? And how do researchers choose which case studies to compare? To examine how planners can learn from systematic case comparison to develop policies, tools, and programs, we turn to the process of policy transfer.

Learning from Successes and Failures Elsewhere

Policy transfer can be described as:

> *the process by which knowledge about policies, administrative arrangements, institutions and ideas in one political system (past or present) is used in the development of policies, administrative arrangements, institutions, and ideas in another political system.* (Dolowitz and Marsh 2000, 5)

Transferable knowledge, in this case, includes policy goals, content, instruments, and programs; institutions, ideologies, and attitudes; and even negative lessons. Although policy transfer can occur at the national level and involve government task forces, politicians and political parties, and international organizations, they can also involve local think tanks, activists, research institutes, and media. Policy transfer is common in planning, and in recent years there has been a growing body of research on the processes, barriers, and outcomes of policy transfer (e.g., McCann 2008; Spaans and Louw 2009; Marsden and Stead 2011; Pojani and Stead 2014, 2018; Lieto 2015; Hurley and Lamker 2016; Monios 2017).

Actors involved in policy transfer processes can learn from the political systems or units within their own country (e.g., broadening provincial solutions to the national level) or from other countries (Dolowitz and Marsh 2000; Spaans and Louw 2009). Marsden and Stead (2011) and Stead (2016) note that local actors tend to look within their own region or country, while national actors tend to look internationally. Policies are much more likely to be transferred between countries that are "psychologically proximate" (Stone 2004; Gurran et al. 2014): those that are geographically, ideologically, or culturally similar. Even though, as we pointed out earlier, there is value in using atypical case studies in analytic generalization (Yin 1994; Flyvbjerg 2001; Schofield 2002), policies are often rejected because the originating context is too dissimilar to the borrowing country, region, or city (Thomas and Bertolini 2014; Stead 2012; Pojani and Stead 2018).

Dolowitz and Marsh (2000) identified four different types of policy transfer: copying (direct and complete transfer), emulation (transfer of the ideas behind the policy or program), combinations (mixtures of different policies), and inspiration (policy in another jurisdiction may inspire a policy change, but the final outcome does not actually draw upon the original). Spaans and Louw (2009) describe different "intensities of transfer" including using policy ideas for inspiration, learning, and transplantation. Akin to learning from case comparison, Stone (2004) argues that we can learn from more than one jurisdiction at a time and take away a few lessons, which would lead to hybrids and adaptive innovation to fit the local context.

Policy transfer occurs for a number of reasons. Policy makers, politicians, or civil servants may be dissatisfied with current policies (Marsden and Stead 2011); they may not have any local examples of solutions to planning problems, so they feel the need to look elsewhere for solutions. For example, McCann highlights the "global circuits of knowledge" (2008, 895) used to develop Vancouver's four-pillar drug strategy. Vancouver's mayor sought out international expertise because the city's deteriorating health and housing conditions in the late 1990s were severely impacting intravenous drug users. The process of policy learning led to the adoption of the first harm reduction strategy of its kind in North America. Uncertainty is also a powerful force that encourages imitation (DiMaggio and Powell 1983); conditions of uncertainty include absence of a scientific consensus, lack of information, new problems, policy disasters, crisis, and political conflict (Stone 1999). In cases of international policy transfer, countries may be encouraged to adopt policies from other places as a condition for a loan or business development, or feel that experts in other countries are more advanced than their own (Pojani and Stead 2018). Stone (2004) contends that governments and international organizations no longer have the ability to design and/or implement effective public policies and that global public policy networks can help. Monios (2017) suggests that in a neoliberal era, the decrease in governmental intervention and diminished role of the government contributes to their lack of expertise in policy development. Spaans and Louw (2009) write that governments are no longer capable of directing planning in the same manner as before, because long drawn-out procedures and long-term visions with a fixed final view no longer seem appropriate; the pace of change is now too quick, demanding new, faster policy development processes. Marsden and Stead (2011) write that policy transfer is becoming more prevalent because of the increased networking opportunities between policy makers and politicians and the pressures of increased accountability.

As Marsden and Stead (2011) found in their review of the literature, there has been little attention paid to policy transfer in transportation. In their study of urban transportation policy transfer in Wroclaw, Poland, and Riga, Latvia, Stead et al. (2008) found that political and administrative stability, the presence of forward-looking policy makers and civil servants, and a small, tight network of participating actors were important factors for successful transfer. In Pojani and Stead's examination of policy transfer to Albania (2018), local transportation plans developed with the aid of foreign expertise were often not implemented due to a lack of coordination in technical planning efforts among the foreign experts, lack of funds to implement the plans, failure to circulate the plans widely to planners and the public, lack of operational details in the plans, ignorance of local planning and social cultures, and lack of education and awareness-raising campaigns parallel to plan development. Monios (2017) found that when policy transfer was influenced by political agendas in transportation planning, one result was "policy churn" (the cyclical adoption of policies, with reforms added, in order to satisfy political objectives rather than to achieve successful implementation).

Policy Transfer in Transit-Oriented Development

If we return to the reasons policy transfer might occur, we can see that many of these apply to planning for TOD. Many regions around the world have adopted TOD as a planning strategy to accommodate growth management and sustainability goals, but most countries have only a few examples of TOD. The process of developing TOD is a long-term process involving governance at different levels, a range of disciplines, and land use modifications, which has meant little implementation in many regions. Lacking their own cases, local actors search for international examples. International exchanges between planners and civil servants are common and TOD best practices show a consistent track record of benchmarking against foreign examples (Tan et al. 2014b). Participants in these exchanges are interested in which policies and practices have been successful, which actors have been involved, what some of the challenges have been, and how things could have been done differently.

The Netherlands illustrates many of the characteristics that would compel local actors to learn from international examples: Dutch transportation planners accept the ideas behind TOD, but have not been able to implement it beyond scattered railway station projects. TOD remains a transportation policy concept and has not been transferred to other fields such as public health or to other stakeholders such as political representatives or property developers. Some railway station areas have been redeveloped through the National Key Projects, and policies have attempted to concentrate growth around rail stations (Bertolini and Spit 1998). But Dutch cities and regions have not yet implemented TOD at the regional scale, which involves a metropolitan-wide network of local developments connected by public transportation (e.g., Knowles 2012; TransLink 2012). Planners involved in TOD in the Netherlands refer to "the believers," those individuals who believe strongly in the implementation of TOD and advocate it to all who will listen (Thomas and Bertolini 2015). These individuals, working in municipal or regional government bodies or as private consultants, have been instrumental in organizing and attending international TOD conferences and exchanges between cities. However, transfer of policy ideas such as the need for collaboration of land use and transportation planners in TOD processes, or for interdisciplinary teams in TOD implementation, seems to be minimal (Tan et al. 2014a).

Concerns About Policy Transfer

There are a number of concerns about taking policy ideas from one location and adopting them in another. Uninformed, incomplete, or inappropriate policy transfer may occur (Dolowitz and Marsh 2000): the borrowing country may have insufficient information about the policy and how it works, crucial elements may not be transferred, or insufficient attention may be paid to the differences in political, institutional, or ideological contexts in the borrowing and transferring countries. Stead (2012, 108) cautions that:

> *Clearly, the less detailed an example of best practice is (and the more sanitized the account of its design or implementation), the less likely it will be that the example can be replicated elsewhere.*

An Organisation for Economic Co-operation and Development report on Best Practices in Urban Development (OECD 2001) characterizes the ways in which best practices are transferred to other countries: some components of the best practice have low transferability (ideas, principles for action, philosophy, programs, institutions, modes of organizations, practitioners, and joint projects), while others have high transferability (methods, techniques, know-how, and operating rules). This suggests that it is often easier for a "borrowing" country to focus on the more technical aspects of a policy than the broader goals, ideals, or institutional frameworks that favored implementation; this narrower focus could contribute to policy failure.

Spaans and Louw (2009) caution that planning problems may manifest themselves in the same way in another country, but the solutions may be restricted to the originating country or the problem may be perceived differently in the borrowing country. Stead (2012) and Pojani and Stead (2018) highlight a number of concerns in asymmetrical policy transfer, e.g., when there are unequal power relationships, resources, and levels of economic development between source and borrower countries. Among policy transfer processes between European countries, these include pressure on a non-EU member country to bring their goals in alignment with EU norms, standards, or instruments, unwillingness to commit to implementing the recommendations of foreign experts, differences in institutional structure (e.g., lack of regional governments), and the inability of foreign experts to recognize cultural differences that would impact plan implementation.

While it may be assumed that planners, politicians, and public servants in the "borrower" country want to seek out external advice and learn from other countries, this is not always the case. Stead (2012) writes that only a minority of officials believes that learning from others plays a large or significant role in their decision making; they tend to value informal contact with peers more. The OECD (2001) suggests that top-down processes of learning from best practices, e.g., those initiated by national or international agencies, may be less successful than those initiated by local actors in response to a need they have recognized themselves. Indeed, in their study of Albanian planners' integration of foreign expertise, Pojani and Stead (2018) found there was low or nonexistent motivation among local planners to implement the advice of experts who had been hired directly by an external funding organization, a common practice in the years immediately following the fall of communism in Albania. They also found that some policies proposed by foreign consultants did not take into account local customs, such as the informal nature of doing business, and that in some cases spatial planning strategies could not be implemented because the required governance structures were not in place in Albania.

Monios (2017) suggests that the neoliberal paradigm in planning means that policy ideas that advance the goals of economic growth are far more likely to be implemented than those that would threaten the existing policy paradigm or those that would be politically unpopular, even if the policy ideas have been proven to be successful in achieving the stated goals. Goldberg-Miller (2018, 173), in her study of municipal cultural policy, points out that:

> *Observational knowledge mobilization, through city case study research, does not question the underlying core beliefs behind the policy solutions, but focuses on the technical implementation of these policies.*

Goldberg-Miller (2018) and Gurran et al. (2014) highlight the similarities between policies and planning practices as a result of policy transfer processes. There is a risk that, despite very different cultural and professional contexts, policy transfer can lead to a "more superficial rather than substantive transfer of rhetorical devices and reform techniques" (Gurran et al. 2014, 196): a convergence of policy without a true consideration of the differences in application across countries.

There are many reasons that we could learn how to implement TOD from other contexts, but how do we do this in a meaningful way that acknowledges these potential concerns?

Meaningful Learning from Other Contexts

Using Case Selection to Shape Learning

Grin (2010) writes that instances of learning may contribute to profound structural change, the type of changes that are needed to break through barriers to implementation. Arguably, in order to achieve these breakthroughs, the choice of case studies is critical. Choosing cases that are not very innovative, for example, may not provide planners and policy makers with the impetus to introduce their own innovative approaches. This has contributed to the belief that only "best practice" cases are worth studying, particularly in TOD. However, in reality there is as much to be gained from learning about "worst practices" as well as the lauded best practice approaches. Stead (2012) suggests that best practices can be problematic because they are often sanitized of political challenges involved in their implementation, or social or cultural issues; sometimes planners, aware that only "good news" stories are disseminated, "pursue their own networks of knowledge in order to gain an understanding of the processes involved" (p. 109). He wrote about policy transfer between East and West European countries:

> *Because of the diversity of member states, institutions, planning instruments, and cultures, it is perhaps more appropriate to consider a move away from the idea of best practice examples and refer instead simply to examples of practice, which policy officials can draw on and adapt to their own circumstances.* (2012, 113)

Monios (2017) and Stead (2012) both suggest that when foreign examples are used, there are no mechanisms to evaluate them to determine which is most suitable or successful. Monios (2017, 360) cautions that in looking elsewhere, policy makers may have their own preferences, so:

> *...seeking and finding a new policy in the first place may not be an exhaustive search, and the lack of good information makes it difficult to validate if it is actually a suitable policy, which leads to a higher likelihood of inappropriate transfer (or indeed incomplete or uninformed). This leads to a higher risk of failure...*

Several authors write about the benefits of learning from negative lessons (Rose 2005; Maxwell 2004; Stone 2004; Thomas and Bertolini 2014). We may also choose cases to learn about different aspects of a policy: Stone (2004) distinguishes between "softer" aspects like ideas, paradigms, interpretations, and problem definitions and "hard" aspects like instruments, legislation, techniques, and policies. When Papa and Bertolini compared six Western European cities to determine the link between TOD and urban form, they deliberately chose cases with "sufficient variation on possible determinants," not potential best or worst practice cases (2015, 73). The US Department of Transportation (2018) chose 13 metropolitan areas with documented evidence of planning for shared mobility, and which represented a range of population size, geography, and land use typographies, to ensure that a diverse set of planning contexts was included.

Decontextualizing and Recontextualizing

As discussed earlier, the differences in cultural, geographic, political, and other characteristics between countries often act as barriers to learning from planning practice and policy development in other countries. One way to approach this is to decontextualize the policy idea from the originating place, presenting it to local practitioners as a concept:

> *Ideas exchanged, negotiated, and translated in cross-border planning networks are not statements of invariant truth, but take the form of myth. In that form, they are often depoliticized and through their depoliticization they can be repoliticized on the ground*... (Lieto 2015, 125)

Although, as many authors have suggested, there is a danger in oversimplifying policy ideas so they be more easily transferred, we suggest that there is value to this approach as part of a learning process. It allows learners to absorb the concept on its own, without considering the unique context that produced it, e.g., the very high density development in Japanese cities which has contributed to corridor development supporting TOD, or the unique institutional role of railway companies as property developers. Decontextualization opens them up to the concept itself, which they might have immediately dismissed otherwise as too foreign or unusual to be considered in their own country. As they learn about and explore the concept, they can better understand its value to their own context by asking questions such as:

- "Would this concept work here? If not, why not?"
- "Do we have a similar governance structure? Would we make changes to ours to make this happen?"
- "How likely would my organization be to adopt this practice?"
- "Do we have a shared understanding of the problem in our city/country?"
- "Would I bring this idea into a discussion about a new policy approach?"

In this way, the local practitioners begin to better understand their own context and its particular processes and whether they would even be open to ideas from other places. They begin to recontextualize the concept into a potential policy idea that is specific to their own place. In a workshop or gaming session (e.g., Straatemeier et al. 2010), they may even go beyond this, with the creative generation of original policies as a result. Lieto (2015) rather poetically suggests that hybrid ideas can be created when each party has its own myth of the policy idea and uses that to develop a new, common myth. This is Dolowitz and Marsh's (2000) third type of policy transfer (combinations): when considering an idea from another place, practitioners in the borrowing country can begin to combine the new idea with their own expertise and understanding of the problem to develop a solution that makes sense in the local context. They may even develop completely new policy concepts as a result of the learning exercise, which aligns with Dolowitz and Marsh's fourth type of policy transfer (inspiration).

Decontextualizing and recontextualizing also makes sense when considering TOD policy, which is very complex. Involving multiple levels of government, funding mechanisms, land use planning, development tools, and collaborative stakeholder engagement practices, TOD policy and implementation often seems to be successful in spite of, rather than because of, an innovative idea.

What, if Anything, Can We Learn from Other Places?

Planning as a discipline draws upon many different fields for its theory, ideas, frameworks, and institutions. Planners regularly study policies from other places, adapting them to suit local governance, policy, and cultural contexts; planners conducting research often turn to the case study. The ability of the case study methodology to integrate multiple methods makes it particularly useful in planning, as a multidisciplinary field integrating a range of methods and approaches. While single-case studies can often provide unique opportunities to focus on the intricacies of specific processes or institutions and are often used to draw broader conclusions, it can be difficult to translate their findings into recommendations for planning practice. Case study comparison can be used to draw out similarities between cases. Cross-case techniques can allow researchers to mine case study data for common elements. Case study comparison, or systematic cross-case techniques such as meta-analysis, can provide planners with more generalizable results than single-case studies.

Policy transfer allows learning about the policy ideas, institutions, and arrangements from another locale. It can include learning from the political systems or units within one's own country, or from other countries. Policy transfer occurs for a number of reasons, a number of which apply to planning for TOD: often, there are few examples of TOD within one's own country, there are established practices of learning from other countries (e.g., international exchanges, networking opportunities), or there is a lack of expertise in policy development. But there are a number of concerns about taking policy ideas from one location and adopting them in another. Some components of policies have low transferability, while others are easier to adopt. Planning problems may manifest themselves in the same way in another country, but the solutions may be restricted to the originating country or the problem may be perceived differently in the borrowing country. Planners, politicians, and public servants in the borrowing country may not want to seek out external advice and learn from other countries; there may be political or other types of pressure to do so. Powerful concepts such as the neoliberal paradigm or the creative cities approach could be transferred to the borrowing country, either unwittingly or not. Policies and planning practices may start to look like pale imitations of each other after a while, until and unless a country establishes its own policy identity.

To combat this, negative lessons, or a range of successes and failures, could be used to understand the constraints and challenges of policy implementation. Decontextualizing a policy idea so that it may be used in a learning exercise might also help local practitioners better understand their own contexts and processes and whether they would be open to ideas from other places. If they are open to learning, developing hybrid or new policy ideas, the lessons from other places could serve merely as inspiration. Learning the concepts, rather than adopting policy ideas altogether, may be a more realistic approach that is less prone to failure in implementation.

References

Andersson, T., Carlsen, J., & Getz, D. (2002). Family business goals in the tourism and hospitality sector: Case studies and cross-case analysis from Australia, Canada, and Sweden. *Family Business Review, 5*(2), 89–106.

Baaijens, S., & Nijkamp, P. (2001). Meta-analytic methods for comparative and exploratory policy research: An application to the assessment of regional tourist multipliers. *Journal of Policy Modeling, 22*(7), 821–858.

Bertolini, L., & Spit, T. (1998). *Cities on rails: The redevelopment of railway station areas.* London/New York: Spon/Routledge.

Bertolini, L., Curtis, C., & Renne, J. (2012). Station area projects in Europe and beyond: Towards transit oriented development? *Built Environment, 38*(1), 31–50.

Cervero, R. (1998). *The transit metropolis: A global inquiry.* Washington, DC: Island Press.

Curtis, C., Renne, J. L., & Bertolini, L. (2009). *Transit oriented development: Making it happen.* Farnham/Burlington: Ashgate Publishing.

Dimaggio, P., & Powell, W. (1983). The iron cage revisited: Institutional isomorphism and collective rationality in organisational fields. *American Sociological Review, 48*, 147–160.

Dixon-Woods, M., Agarwal, S., Jones, D., Young, B., & Sutton, A. (2005). Synthesising qualitative and quantitative evidence: A review of possible methods. *Journal of Health Services Research and Policy, 10*(1), 45–53.

Dolowitz, D., & Marsh, D. (2000). Learning from abroad: The role of policy transfer in contemporary policy making. *Governance, 13*(1), 5–24.

Ewing, R., and Cervero, R. (2010). *Effects of TOD on housing, parking, and travel.* Transportation research board: Transit cooperative research program report 128.

Flyvbjerg, B. (2001). *Making social science matter: Why social inquiry fails and how it can succeed again.* Cambridge: Cambridge University Press.

Flyvbjerg, B. (2006). Five misunderstandings about case study research. *Qualitative Inquiry, 12*(2), 219–245.

Goldberg-Miller, S. B. D. (2018). Keeping creativity downtown: Policy learning from San Francisco, Seattle, and Vancouver for municipal cultural planning in Toronto. *The Journal of Arts Management, Law, and Society, 48*(3), 170–190. https://doi.org/10.1080/10632921.2017.1422834.

Grin, J. (2010). The governance of transitions. In J. Grin, J. Rotmans, & J. Schot (Eds.), *Transitions to sustainable development: New directions in the study of long term transformative change* (pp. 265–285). New York and London: Routledge.

Gurran, N., Austin, P., & Whitehead, C. (2014). That sounds familiar! A decade of planning reform in Australia, England, and New Zealand. *Australian Planner, 51*(2), 186–198.

Hurley, J., & Lamker, C. W. (2016). Exchange between researchers and practitioners in urban planning: Achievable objective or a bridge too far? *Planning Theory & Practice, 17*(3), 447–453. https://doi.org/10.1080/14649357.2016.1190491.

Khan, S., & Van Wynsberghe, R. (2008). Cultivating the under-mined: Cross-case analysis as knowledge mobilization. *Forum: Qualitative Social Research, 9*(1), article 34. http://www.qualitative-research.net/index.php/fqs/article/view/334/729

Knowles, R. D. (2012). Transit oriented development in Copenhagen, Denmark: From the finger plan to ørestad. *Journal of Transport Geography, 22*, 251–261.

Lieto, L. (2015). Cross-border mythologies: The problem with traveling planning ideas. *Planning Theory, 14*(2), 115–129.

Marsden, G., & Stead, D. (2011). Policy transfer and learning in the field of transport: A review of concepts and evidence. *Transport Policy, 19*, 492–500.

Maxwell, J. A. (2004). Using qualitative methods for causal explanation. *Field Methods, 16*(3), 243–264.

McCann, E. J. (2008). Expertise, truth, and urban policy mobilities: Global circuits of knowledge in the development of Vancouver, Canada's 'four pillar' drug strategy. *Environment and Planning A, 40*, 885–904. https://doi.org/10.1068/a38456.

Miles, M. B., & Huberman, A. M. (1994). *Qualitative data analysis: An expanded sourcebook*. Thousand Oaks: Sage.

Monios, J. (2017). Policy transfer or policy churn? Institutional isomorphism and neoliberal convergence in the transport sector. *Environment and Planning A, 49*(20), 351–371.

Naess, P., Strand, A., Naess, T., & Nicolaisen, M. (2011). On their road to sustainability? The challenge of sustainable mobility in urban planning and development in two Scandinavian capital regions. *Town Planning Review, 82*(3), 285–315.

Newman, P. (2007). Planning for TOD in Australian cities. *Building Environment Design Professionals Environment Design Guide, 2*(15), 1–11.

Nijkamp, P., van der Burch, M., & Vindigni, G. (2002). A comparative institutional evaluation of public-private partnerships in Dutch urban land-use and revitalization projects. *Urban Studies, 39*, 1865–1880.

OECD. (2001). *Best practices in local development*. Paris: OECD.

Papa, E., & Bertolini, L. (2015). Accessibility and transit-oriented development in European metropolitan areas. *Transport Geography, 47*, 70–83.

Pojani, D., & Stead, D. (2014). Ideas, interests, and institutions: Explaining Dutch transit oriented development challenges. *Environment and Planning A, 46*, 2401–2418. https://doi.org/10.1068/a130169p.

Pojani, D., & Stead, D. (2018). When west–east planning policy advice fails to gain traction. *Journal of Environmental Planning and Management*. https://doi.org/10.1080/09640568.2018.1497586.

Ragin, C. C. (1987). *The comparative method: Moving beyond qualitative and quantitative strategies*. Berkeley: University of California Press.

Rose, R. (2005). *Learning from comparative public policy: A practical guide*. London and New York: Routledge.

Sandelowski, M., Cocherty, S., & Emden, C. (1997). Qualitative metasynthesis: Issues and techniques. *Research in Nursing and Health, 20*, 365–371.

Schofield, J. W. (2002). Increasing the generalizability of qualitative research. In A. M. Huberman & M. B. Miles (Eds.), *The qualitative researcher's companion* (pp. 171–204). Thousand Oaks: Sage.

Spaans, M., & Louw, E. (2009). *Crossing borders with planners and developers and the limits of lesson-drawing. City futures in a globalising world*. Madrid: University Rey Juan Carlos of Madrid.

Stead, D. (2012). Best practices and policy transfer in spatial planning. *Planning Practice and Research, 27*(1), 103–116.

Stead, D. (2016). The use of academic research in planning practice: Who, what, where, when and how? *Planning Theory & Practice, 17*(3), 453–456. https://doi.org/10.1080/14649357.2016.1190491.

Stead, D., Dejong, M., & Reinholde, I. (2008). Urban transport policy transfer in central and Eastern Europe. *Disp: The Planning Review, 44*, 62–73.

Stone, D. (1999). Learning lessons and transferring policy across time, space and disciplines. *Politics, 19*, 51–59.

Stone, D. (2004). Transfer agents and global networks in the "transnationalisation" of policy. *Journal of European Public Policy, 11*, 545–566.

Straatemeier, T., Bertolini, L., te Brömmelstroet, M., & Hoetjes, P. (2010). An experiential approach to research in planning. *Environment and Planning B, 37*, 578–591.

Tan, W. (2011). *NICIS KEI case study #1: Perth, Western Australia*.

Tan, W. G. Z., Janssen-Jansen, L. B., & Bertolini, L. (2014a). The role of incentives in implementing successful transit-oriented development strategies. *Urban Policy and Research, 32*, 33–51.

Tan, W. G. Z., Janssen-Jansen, L. B., & Bertolini, L. (2014b). Identifying and conceptualizing context-specific barriers to transit-oriented development strategies: The case of the Netherlands. *Town Planning Review, 85*, 639–663.

Thomas, R., & Bertolini, L. (2014). Beyond the case study dilemma in urban planning: Using a meta-matrix to distil critical success factors in transit-oriented development. *Urban Policy and Research, 32*(2), 219–237. https://doi.org/10.1080/08111146.2014.882256.

Thomas, R., & Bertolini, L. (2015). Policy transfer among planners in transit-oriented development. *Town Planning Review, 86*(5), 537–560. https://doi.org/10.3828/tpr.2015.32.

Translink. (2012). *Transit-oriented communities guidelines. Creating more livable places around transit in metro Vancouver*. Vancouver: TransLink.

United States Department of Transportation (USDOT). (2018). *Integrating shared mobility into multimodal transportation planning: Improving regional performance to meet public goals.* Report number DOT-VNTSC-FHWA-18-13.

Van Egmond, P., Nijkamp, P., & Vindigni, G. (2003). A comparative analysis of the performance of urban public transport systems in Europe. *International Social Science Journal, 55*(176), 235–247.

Walter, A. I., & Scholz, R. W. (2007). Critical success conditions of collaborative methods: A comparative evaluation of transport planning projects. *Transportation, 34*, 195–212.

Yin, R. K. (1981). The case study crisis: Some answers. *Administrative Science Quarterly, 26*(1), 58–68.

Yin, R. K. (1994). Case study research: Design and methods (2nd ed.). Thousand Oaks, CA: Sage.

Yin, R. K., & Heald, K. A. (1975). Using the case survey method to analyze policy studies. *Administrative Science Quarterly, 20*, 371–381.

CHAPTER 3

International Case Studies in TOD

Abstract Many cities and regions around the world would like to implement TOD. Planners, politicians, civil servants, and others can use policy ideas from other places to catalyze policy development in their own city-region, if they are careful to use them as learning and inspiration rather than copying them. In this chapter, we present the 11 case studies included in our study of transit-oriented development implementation, including brief profiles of each city-region and a meta-analysis. We present the factors that have been critical in the successful implementation of TOD, along with some examples of policy ideas/lessons that increased success in the case studies. Finally, we discuss how we used these research findings in a policy learning workshop for planners and policy makers in two Dutch city-regions. We found that the country's unique culture, planning practice, and institutions inevitably shape the barriers and solutions to TOD implementation.

Keywords Policy learning • Collaboration • Actor relationships • Public participation

Many cities and regions around the world would like to implement TOD for its benefits, including decreased driving, parking needs, and congestion; increased transit ridership, mobility choice, and access to jobs and services. We know that planners, politicians, civil servants, and other

R. Thomas, L. Bertolini, *Transit-Oriented Development*,
https://doi.org/10.1007/978-3-030-48470-5_3

stakeholders can learn from other places, using policy ideas to catalyze policy development in their own city-region, if they are careful to use them as learning and inspiration rather than copying them.

In this chapter, we present the 11 case studies included in our study of transit-oriented development implementation. We discuss how we chose the case studies, present profiles of each city-region, and discuss results from a meta-analysis of the case studies. We present the factors that have been critical in the successful implementation of TOD, along with some examples of policy ideas/lessons that increased success in the case studies. Finally, we discuss how we used these research findings in a policy learning workshop for planners and policy makers in two Dutch city-regions. The crux of our research study is trying to understand whether local planners can use policy ideas/lessons as learning or inspiration to develop their own solutions to TOD implementation. The goal was to catalyze new ways of thinking to force a breakthrough in Dutch city-regions, which have been struggling to implement TOD beyond single railway station areas, as a comprehensive set of policy tools to address issues such as sustainability, growth management, and public health.

While our results are certainly useful for other cities and regions seeking to implement TOD in a comprehensive way, we found that barriers can persist. A country's unique culture, planning practice, and institutions inevitably shape the barriers and solutions to TOD implementation.

Choosing the Case Studies

As discussed in Chap. 2, there have been some previous attempts to compare cases in TOD: Bertolini and Spit (1998), Cervero (1998), Newman (2007), Curtis et al. (2009), and Bertolini et al. (2012) each attempted to identify some commonalities among the cases. However, case comparison was not their sole concern; therefore, they did not devote much time to exploring the common elements or characteristics of TOD, or factors contributing to successful implementation across cases.

For this study, one of our goals from the outset was to go beyond the single-case study dilemma, synthesizing existing case studies to distill generalizable patterns across the cases. Beginning with a long list of over 30 potential international city-regions, we chose case studies that met the following criteria:

- There were completed case studies, in the form of case study reports, policy documents, and supporting academic literature, on TOD in the city-region.
- The city-region had been attempting to integrate transportation and land use policy through TOD implementation for at least 20 years.
- Three local experts in each city were available to assist the authors with the research.

The cases that met these criteria were Tokyo, Perth, Melbourne, Montreal, Vancouver, Toronto, Naples, Copenhagen, Amsterdam-Utrecht, Rotterdam-The Hague, and Arnhem-Nijmegen.

As discussed in Chap. 2, a range of "successful" and "unsuccessful" cases, according to the definition of TOD we developed from the literature, was considered essential. Knowles (2012) and TransLink (2012) both characterized TOD as part of a broader smart growth approach to urban development and growth management. The TOD implementation approach in European cities also began evolving in the 2000s to include the more regional approach to travel demand seen in American and Australian cities, rather than concentrating on the economic revitalization and multimodal redesign of individual railway station areas (Bertolini et al. 2012). Therefore, for a case to be considered successful in this study, the city-region had to be attempting to shift travel mode choice from driving to walking, cycling, and transit at the regional scale. The (re)development of single, dispersed railway stations was not considered to be as successful as (re)development of linear corridors supporting walking, cycling, and transit (Chorus and Bertolini 2016). Based on these considerations, our definition of TOD is as follows:

> *TOD can be described as land use and transportation planning that makes walking, cycling, and transit use convenient and desirable, and that maximizes the efficiency of existing public transit services by focusing development around public transit stations, stops, and exchanges. Successful TOD can be defined as implementation of this type of development at a regional scale.*

Analyzing the case study documents included creating a report for each city-region, then coding the case reports using the five major themes found in the literature:

- Policy consistency
- Actors and their roles
- Connections between land use and transportation
- Specific policies and instruments
- Barriers to TOD

The coded data was then entered into a meta-matrix: a spreadsheet that enabled us to easily identify patterns across the cases. Before we discuss the process in more detail, we will introduce the case studies.

City-Region Profiles

This section profiles each city-region, focusing on the five themes and illustrating some of the major strengths and weaknesses in each case.

Tokyo

The Tokyo case shows very strong coordination of land use and transportation planning policy and consistent policy for several decades supporting the development of urban corridors supporting transit use (Chorus 2012). Railway companies in the region own and develop land along the routes, with a particular focus on attracting users all day. With the nation's vast network of railways and high urban densities, there is a high level of national government support for TOD. Municipal master plans involve extensive consultation with private-sector developers, especially the property division of railway companies, sometimes at the expense of public participation in this process. Building heights and footprints are often unpredictable, due to flexible planning guidelines that allow developers to apply for floor area ratio (FAR) bonuses and transfers to achieve higher densities near railway stations or redevelopment districts specified by municipal governments.

Perth

Key visionaries have been critical in promoting the TOD concept in Perth, which has had inconsistent land use policy and no vision for the region until recently (Curtis 2008; Tan 2011). A vision for the region began to take shape during extensive public participation, including efforts such as Dialogue with the City and TravelSmart. There is a high level of willingness to experiment with new policies: Perth has implemented several TOD projects (Mouritz and Ainsworth 2009). There has been some increased collaboration among actors involved in TOD, but formal and informal changes to land use and transportation institutions have been unstable as higher-level governments have changed; a fair amount of fragmentation of governance related to TOD still exists. Differing definitions of TOD among transportation and land use planners complicate implementation.

Melbourne

Considerable fragmentation of governance at the municipal and regional levels and a low level of communication and coordination have contributed to distinct and separate land use and transportation policies in Melbourne (Legacy et al. 2012). There is not yet a consistent vision for future growth and development, although public acceptance of public transit and higher densities is on the rise. Although there is no regional land use planning authority, informal collaboration among local government associations has been integrated in the development of innovative strategies and plans. Some willingness to experiment can be seen in a 2008 forum on good governance for transportation infrastructure, where stakeholders discussed institutional integration and barriers to implementation. Some transit-accessible activity centers have been designated, but many are in peripheral locations (Curtis and Scheurer 2012).

Montreal

Transportation and land use planning remain quite separate in Montreal, with little collaboration between departments on plans and policies. There is some consistent land use planning policy, such as the use of density bonuses, long-term leases, and FAR exceptions, which are used to concentrate growth around metro stations (El-Geneidy et al. 2011). Governance in the region is highly fragmented, involving a large number of actors, and provincial government support for transportation infrastructure is unstable. A number of TOD projects have been built (Adenot 2007a, b), though they fit the pattern of single and dispersed, rather than following a linear or networked pattern as part of a regional strategy.

Vancouver

Vancouver has had a very consistent vision for decades (first the Livable Region Strategic Plan, then the Livable Region Plan, and now the Regional Growth Strategy) and high political stability at the local level. The situation at the provincial and national level has been less stable, contributing to inconsistent support for TOD. Implementation has also been limited by strong inter-municipal competition for development projects and funding (Raad and Kenworthy 1998). The advisory regional land use planning authority works closely with the regional transportation authority to develop regional plans and guidelines and designate transit corridors for both Bus Rapid Transit (BRT) and Light Rapid Transit (LRT) (TransLink 2012). Municipal land use plans in the region integrate the Frequent Transit Network developed by the regional transportation authority in order to plan for new growth. The City of Vancouver clearly describes the use of tools such as density bonuses and FAR transfers on its website, so that developers understand how and where they can be used. Public participation in planning processes is very high and educational efforts have increased support for TOD (Tan et al. 2012).

Toronto

The City of Toronto has supported higher densities around metro stations through the use of zoning incentives for many years. Decades of inconsistent growth policies in Toronto have resulted in no regional vision, despite Toronto's inclusion in a larger regional growth plan encompassing the mega-region (containing 110 municipalities) led by the province. Lack of federal and provincial support for transportation infrastructure, a high level of instability in local politics, and strong inter-municipal competition for development project also complicate TOD implementation. The first regional transportation strategy was developed in 2006 on the initiative of Toronto's first regional transportation authority; this has somewhat diminished the formerly high level of fragmentation among transportation providers in the region, though metropolitan governance is still quite fragmented. There are several high-density corridors with high access to public transportation and mixed-use centers even in some suburban areas (Filion 2001; Filion et al. 2006), but TOD at a regional scale is limited. Public support for higher densities and TOD and public involvement in transportation processes are much lower than would be expected in such a large metropolitan area (Perl and Pucher 1995; Grant 2002; Filion and McSpurren 2007).

Naples

Naples was a particularly interesting case: political stability at both the local and national level has been an ongoing issue for decades and the city struggled with the approval and implementation of land use planning legislation (Allum 2003; Gelli 2001). These issues contributed to serious issues such as landslides from overdevelopment in vulnerable areas. Intense competition and little collaboration among the actors involved in TOD were also key challenges (Bull 2005). Major changes were initiated in 1993 with the election of Mayor Antonio Bassolino, the first of several key visionaries: small-scale improvements were implemented, such as closing spaces that had been used for parking and reopening them as pedestrian-only public squares. The mayor also attracted the support of higher-level governments, including the European Union, to redevelop key railway stations. After the implementation of a major regional initiative, the Regional Metro System Project of Naples and Campania (Cascetta and Pagliara 2008; Pagliara and Papa 2011), Naples now has widespread implementation of TOD projects, as well as improved collaboration between land use and transportation planners and a multidisciplinary approach to TOD implementation, including architects, developers, engineers, and planners.

Copenhagen

Copenhagen's Finger Plan is an internationally known vision, and one of the earliest examples of TOD (Chapple and Loukaitou-Sideris 2019). However, there has been less implementation than expected because cities have not closely adhered to it, or to other TOD policies (Hartoft-Nielsen 2007; Naess et al. 2011). As a result of local government reforms in 2007 and a national directive on the Finger Plan, there has been more concentrated development around transit stations and in some areas, such as Ørestad. Key visionaries such as Jan Gehl have been active in promoting concentrated development, and public support for inner city revitalization projects and cycling has increased. However, most residents still prefer to live in suburban areas and drive for most of their daily needs (Naess 2005). Remaining challenges to TOD include strong inter-municipal competition for development, lack of coordination among the various actors, and lack of site-specific tools to promote development in the designated areas.

Amsterdam-Utrecht

Although Amsterdam and Utrecht comprise the strategically important North Wing of the Randstad region in terms of transportation (the South Wing includes Rotterdam and The Hague), it does not have a developed regional vision. But as a polycentric city-region, Amsterdam-Utrecht has seen some policy consistency and a high level of government support for TOD. A long history of informal collaboration between the municipalities has not helped them coordinate on TOD implementation, and projects have struggled to move from planning to implementation (Bertolini 1998; Tan 2009). However, cycling policies and programs have significantly shifted travel mode choice away from the car (Schwanen et al. 2004). Minimal public participation in transportation planning, no key visionaries, and a low degree of willingness to experiment with policies or tools are still barriers to TOD implementation.

Rotterdam-The Hague

Like Amsterdam-Utrecht, Rotterdam-The Hague has seen some policy consistency and a high level of government support for TOD. More formal collaboration through administrative platforms (Metropoolregio and South Wing Administrative Platform) ensures that strategies and policies are discussed between the municipalities and city-regions (*stadsregios*). Examples of regional projects requiring extensive collaboration include Randstad Rail, which connects the region's city centers and new housing areas by rail; Stedenbaan, an agreement which allowed 11 jurisdictions to coordinate transportation and land use in the region; and Programmabureau StedenbaanPlus (2011), which expanded both the number of municipal partners (increased to 47) and the scope of TOD (which now includes the quality of station areas). The Stedenbaan agreement resulted in 45 percent of new housing in the South Wing being built near the designated stations from 2006 to 2010 (Keurs et al. 2012). This regional-scale attempt at TOD implementation, and high degree of willingness to experiment with governance and actor relationships, is unique in the Netherlands (Duffhues 2010; Meijers et al. 2012).

Arnhem-Nijmegen

Arnhem-Nijmegen has also seen some policy consistency and a high level of government support for TOD. Images of the historic tram have been used in the development of a new vision for LRT. Informal collaboration between the municipalities includes the development of Stadsregiorail Arnhem-Nijmegen (e.g., frequent commuter train services between the two cities, a BRT line). Transportation planning does not involve much public participation because of the formal negotiation process between municipalities and regional and national authorities (Langendijk and Boertjes 2012). Key visionaries, including a Member of Parliament, have advocated for TOD.

It is clear from these profiles that each of the case studies has their own strengths and weaknesses: none is a complete success or a complete failure at implementing TOD. In order to understand more about the patterns across the cases, we turned to a meta-analysis using a meta-matrix.

Meta-analysis: Decontextualizing

Analyzing the case study documents included creating a report for each city-region, then coding the case reports using five broad codes: policy consistency, actors and their roles, connections between land use and transportation, specific policies and instruments, and barriers to TOD. As the coded data was entered into the meta-matrix, a large table with the case studies along the top and the five broad codes along the side, the patterns became clear.

There were cross-case patterns, related to our definition of success, which persisted across geographical or political differences. As we scanned across the columns of each case, the factors that either facilitated TOD implementation or acted as a barrier to it were noted at the bottom of each column. These "critical success factors" (CSFs), as they are described in the literature (Nijkamp et al. 2002; Van Egmond et al. 2003), were developed through the iterative process of entering the coded data into the meta-matrix and re-ordering the data in chronological order to understand the progress of key plans and initiatives in each case (see Fig. 3.1). This process was highly conducive to learning the data and cases thoroughly, since the qualitative data was read, coded, and then coded data was read again while reordering it within the matrix; this greatly enhanced the researchers' ability to identify patterns.

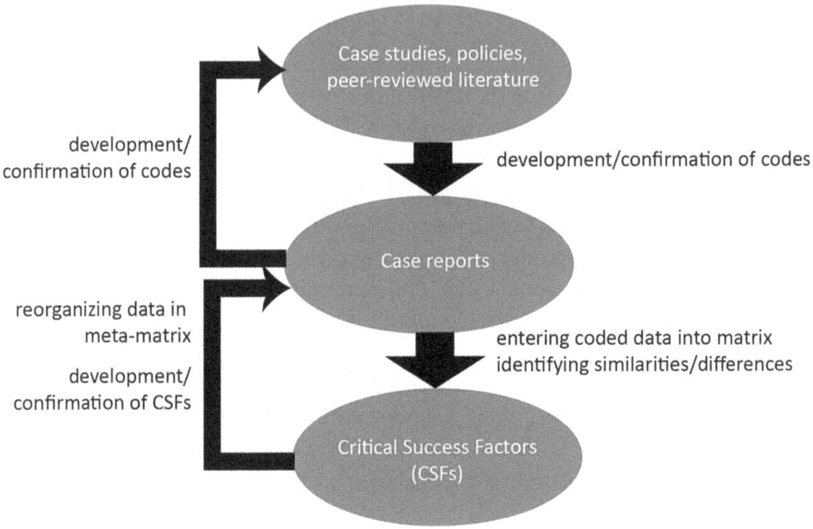

Fig. 3.1 The process of developing the critical success factors (CSFs)

Critical Success Factors

Given the goal of our study to explore actors, policies, and governance arrangements in TOD, as outlined at the end of Chap. 1, it is not surprising that the 16 CSFs fall under three headings: Plans and Policies, Actors, and Implementation. Many of the CSFs are well known to transportation researchers specializing in TOD (e.g., key visionaries, policy consistency), although some are discussed more among practicing rather academic planners (e.g., inter-municipal competition, certainty for developers). Some of the CSFs are entirely under the control of municipal/regional governments, such as public participation in planning processes or policy consistency, while others are not, such as the support of higher-level governments or political stability at the national level. Nevertheless, all 16 factors were present in each case: this approaches the more generalizable results we were aiming for with the meta-analysis of completed case studies, as compared to a single-case study. By taking this approach of cross-case analysis, we believe that we were able to successfully identify the factors that would be instrumental in successful TOD implementation in other city-regions.

The complete list of 16 CSFs is shown in Fig. 3.2.

	WHAT INCREASES SUCCESS?	CRITICAL SUCCESS FACTORS	WHAT DECREASES SUCCESS?
POLICIES &PLANS	Very consistent over time in planning policy supporting TOD (e.g. Copenhagen)	POLICY CONSISTENCY	Very inconsistent policy supporting TOD, major changes
POLICIES &PLANS	Very stable vision, e.g. city-regional vision for land use-transport planning or urban sustainability (e.g. Vancouver, Copenhagen)	VISION STABILITY	Very unstable vision, major changes over time
POLICIES &PLANS	Very good support of higher levels of government (e.g. Naples)	GOVERNMENT SUPPORT	No support of higher levels of government, no policies or funding
POLICIES &PLANS	Very stable national political agenda supporting TOD (e.g. Copenhagen, Tokyo)	POLITICAL STABILITY (NATIONAL)	Very unstable national political agenda supporting TOD, major changes over time
POLICIES &PLANS	Very stable local (municipal or regional) political agenda supporting TOD (e.g. Vancouver, Copenhagen)	POLITICAL STABILITY (LOCAL)	Very unstable local (municipal or regional) political agenda supporting TOD, major changes over time
ACTORS	Very good relationships between municipal actors at a regional scale (e.g. Rotterdam-Den Haag)	ACTOR RELATIONSHIPS	Poor or no relationships between municipal actors at a regional scale
ACTORS	Presence of a regulatory regional land use-transportation planning body (e.g. Perth, Vancouver)	REGIONAL LAND USE-TRANSPORTATION BODY	No regional land use-transportation planning body (advisory or regulatory)
ACTORS	No competition among municipalities for new developments/funding (e.g. Perth, Tokyo)	INTER-MUNICIPAL COMPETITION	Very intense competition among municipalities for new developments/ funding
ACTORS	Widespread presence of multidisciplinary teams implementing TOD (e.g. Naples)	MULTIDISCIPLINARY IMPLEMENTATION TEAMS	Sector-specific teams (e.g. solely planners or engineers) implementing TOD
ACTORS	Very high public participation in land use-transport planning processes (e.g. Perth, Vancouver)	PUBLIC PARTICIPATION	No public participation, public not engaged or interested
ACTORS	Very high public acceptance of high densities and public transit (e.g. Vancouver, Arnhem-Nijmegen)	PUBLIC ACCEPTANCE	No public acceptance of high densities and public transit
ACTORS	Many influential key visionaries over time (e.g. Naples)	KEY VISIONARIES	No key visionaries over time
IMPLEMENTATION	Widespread use of site-specific planning tools (e.g. Tokyo, Vancouver, Montreal, Toronto)	SITE-SPECIFIC PLANNING TOOLS	No use of site-specific tools
IMPLEMENTATION	Corridor-level planning, e.g. coordination of land use and transport in widespread transit corridors (e.g. Tokyo, Vancouver, Toronto)	REGIONAL-LEVEL TOD PLANNING	No corridor-level or station-area planning
IMPLEMENTATION	High degree of certainty for developers (e.g. Tokyo, Perth)	CERTAINTY FOR DEVELOPERS	Uncertainty; developers are unaware of policies, tools and sites encouraging TOD
IMPLEMENTATION	Actors are very willing to experiment with new policies, practices, and tools (e.g. Tokyo, Perth, Arnhem-Nijmege, Rotterdam-Den Haag)	WILLINGNESS TO EXPERIMENT	Actors are unwilling to experiment with new policies, practices, and tools

Fig. 3.2 Critical success factors (CSFs) with case examples showing what increases success

Interpreting the CSFs as suggestions for municipal or regional governments, rather than a list of items to check off when attempting to implement TOD, could help local actors break out of familiar patterns. For example:

- If it has been challenging to develop more clear tools and identify locations where development should occur near transportation nodes, a regional government could develop a list of nodes that should be top priority for new housing or commercial space. They could then help facilitate development in these areas with site-specific tools such as FAR bonuses.
- If neighborhood residents do not support increased densities around transportation corridors, an education campaign could be developed illustrating the need for sites to accommodate population growth in the next decade or two.
- If there is no regional transportation-land use organization, the regional and municipal governments, transportation authorities, and developers could form a working group or steering committee to meet regularly to share methods, approaches, and outcomes from their own initiatives. In time this could become an informal or formal partnership.

In order to assist local actors in understanding the strengths and weaknesses in their own cities, a simple five-point ranking scale could be used (see Fig. 3.3). Here, representatives from the different institutions and organizations could organize a session where they would each rank their city-region on the 16 CSFs. This would help them understand which factors were the weakest: these weaknesses could define where they might want to focus their attention in the coming months or years.

We applied the ranking scale to the case studies and enlisted the help of three experts in each city-region to validate the rankings. We asked them to rank their own city-region using the five-point scale. Then we used these rankings to produce Table 3.1, which shows the aggregate rankings, weighted at 50 percent our scores as researchers and 50 percent local expert scores. Thus, the table shows aggregated rankings for each case study, which reflect not only our interpretation of the completed case study reports and documentation but also the perceptions of professionals in planning practice in the local areas. Once we had these cross-case patterns, we were ready to share our findings with Dutch planners.

CRITICAL SUCCESS FACTORS	SCALE 1 → 5
How consistent would you say planning policy has been in supporting TOD over time? (including specific station areas, transit corridors, cycling and pedestrian infrastructure)	Increasing consistency →
How stable would you say the vision of the city-region has been over time? (e.g. city-regional vision for land use-transport planning or urban sustainability)	Increasing stability →
How much support is there from higher levels of government for TOD? (e.g. state tax on gasoline to support public transit, national station location or regeneration policy, provincial funding for cycling infrastructure)	Increasing support, policies, funding →
How stable has the political situation been at the national level?	Increasing stability →
How stable has the political situation been at the local level?	Increasing stability →
How would you characterize the relationships between actors (public, private, and other) involved in TOD? (e.g. communication, collaboration, overlap in goals or vision, clarity of roles)	Increasing communication, collaboration, overlap on goals →
Is there a regional land use-transportation body/organization (advisory or regulatory)?	Increasing formality, regulatory power →
How much do municipalities in the region compete, rather than collaborate, in terms of new developments / funding?	Decreasing competition, equitable distribution of development →
Are interdisciplinary teams involved in TOD implementation?	Increasing interdisciplinarity →
How would you describe the level of public participation in land use-transportation planning processes?	Increasing participation, engagement →
How would you describe the level of public acceptance of TOD? (including higher densities and public transit in general)	Increasing acceptance →
Have there been key visionaries over time who have influenced the adoption of TOD? (e.g. elected, citizen, or business leaders)	Increasing influence →
Are site-specific planning tools used to support TOD? (e.g. Floor Area Ratio bonuses, leasing of air rights, density targets)	Increasing use of tools across region →
Would you say that TOD is designated, and land use and transport coordinated, at the regional level? (e.g. specific streets as public transit corridors, regional railway corridors)	Increasing coordination of planning in designated corridors →
How would you describe the level of certainty for TOD developers? (e.g. plans and policies supporting higher densities, tools to enable mixed uses at station areas, designation of areas for development / transit corridors)	Increasing certainty, awareness of policies, tools, sites →
How would you describe the willingness to experiment of the actors involved in TOD? (e.g. learning from other contexts, developing new tools)	Increasing willingness →

Fig. 3.3 Critical success factors ranking scale used to determine the strengths and weaknesses of each case

Table 3.1 Codified data matrix

Critical success factors	*TOK*	*PER*	*MEL*	*MON*	*VAN*	*TOR*	*NAP*	*COP*	*AMS*	*ARN*	*ROT*
Plans and Policies											
1 Policy consistency	4	3	3	3	5	2	2	4	3	3	3
2 Vision stability	4	3	2	4	5	3	2	4	3	3	3
3 Government support	3	4	3	3	3	2	4	3	4	4	3
4 Political stability (national)	4	4	3	3	3	3	2	4	2	2	2
5 Political stability (local)	4	4	3	3	4	2	2	5	2	3	3
Actors											
6 Actor relationships	4	3	2	3	4	3	2	3	2	4	4
7 Regional land use-transport body	3	4	1	3	4	3	2	2	1	4	4
8 Inter-municipal competition	4	4	3	1	2	2	2	2	2	3	3
9 Interdisciplinary implementation teams	4	3	2	3	4	3	4	3	3	3	3
10 Public participation	2	4	3	3	5	3	2	2	1	3	2
11 Public acceptance	3	3	2	3	3	3	3	3	3	4	2
12 Key visionaries	2	4	1	4	4	3	3	2	2	3	3
Implementation											
13 Site-specific tools	5	3	2	4	5	3	2	3	1	2	2
14 Regional TOD planning	4	4	2	3	5	4	4	3	1	4	3
15 Certainty for developers	5	4	2	3	4	3	3	3	2	3	3
16 Willingness to experiment	5	4	2	4	4	3	3	3	2	4	4

To see whether local actors in the Netherlands could use ideas from the international case studies to strengthen the weaknesses in their own city-regions, we held two policy learning workshops. These were held in Amsterdam-Utrecht and Rotterdam-The Hague. In order to maximize the potential for planners in the local (Dutch) cases to learn from the other case studies, we deliberately left out the names of the case studies when we gave examples of policy ideas that had worked elsewhere. This "decontextualization" of the policy ideas helped the participants understand the CSFs and policy ideas as universal and the examples as if they could be used in any local context. These policy ideas or lessons could then be understood on their own, e.g., a media campaign, workshops, and gaming to build a vision for the region, instead of attached to their specific context. We considered decontextualization to be necessary because previous research and advocacy on TOD in the Netherlands had shown that Dutch planners were quite averse to considering ideas from cities or countries that were markedly different from their own cultural, geographic, or political contexts, e.g., Tokyo. During the workshops, the stakeholders would begin to recontextualize the ideas to their own local context.

Policy Learning: Recontextualizing

Since TOD, and transportation planning in general, is a complex process involving different levels of government, private- and public-sector organizations, and a variety of administrative arrangements, we wanted to include as many diverse actors as possible in the two workshops. Our goal was to get the participants to use the policy lessons from the other case studies as inspiration or learning, recognizing that complete copying of policy ideas is usually undesirable (e.g., Dolowitz and Marsh 2000; Spaans and Louw 2009). We hoped that the decontextualization of the policy ideas/lessons would help in this process, so that participants could understand what types of solutions might work in the Netherlands, in effect developing solutions that would be tailor-made to the Dutch context. To measure whether the policy ideas/lessons were actually transferred over to the Netherlands, we conducted pre- and post-workshop surveys with the participants.

We designed the workshop based on Grin's (2010) approach to learning. He writes that first-order learning includes helping actors reflect on their own daily actions, within the boundaries of their basic convictions. But second-order learning takes actors beyond their basic convictions, engaging them in a practice that allows them to confront new views that are not normally expressed in their own daily actions. Grin maintains that radical change needs second-order learning; conversely, structural change may promote second-order learning. In order to get Dutch planners out of their regular ways of thinking, we organized:

- An introduction to the CSFs (Fig. 3.2), including the ranking table for all 11 case studies (Table 3.1)
- Application of the five-point CSF ranking scale (Fig. 3.3) to allow participants to understand the strengths and weaknesses in their city-region
- Application of the international policy lessons to the local weaknesses they had identified
- A discussion about the challenges and opportunities of policy transfer to the Netherlands

The two workshops were held in the "North Wing" (Amsterdam-Utrecht) and "South Wing" (Rotterdam-The Hague) of the Randstad region. Participants included urban planners (8), transportation planners (5), architects (2), and project managers (1); some identified with multiple roles. Six researchers in land use and transportation planning from the University of Amsterdam, Delft University of Technology, and Radboud University, the three research project leads, also participated. Participants were sent the list of CSFs beforehand, as well as explanation of findings from the earlier stages of the research and a description of the two exercises that would occur during the workshop.

Identifying Weaknesses

Using the five-point ranking scale, participants identified the following weaknesses in the Amsterdam-Utrecht and Rotterdam-The Hague regions:

- Weak relationships with actors outside of the planning discipline and particularly with the national government
- An unwillingness to experiment with new administrative arrangements, policies, or ideas
- A lack of public participation, which contributed to little knowledge about planning in the broader community

Weak Actor Relationships

There is a very small group of TOD practitioners in the Netherlands who are well-versed in the technical aspects of TOD, such as infrastructure planning. Although they meet fairly regularly and organize events, as one architect said, "We've been talking about this for… 30 years now. We should know what to do now." An urban planner in the Amsterdam workshop noted the "lack of trust in the parties that are involved in development and implementation."

Stronger actor relationships emerged in The Hague workshop: participants mentioned Stedenbaan, the formal agreement between municipalities on strategic transportation and land use planning between Rotterdam, The Hague, and nine other municipalities. But even here, participants noted that there was too much talk and too little action: consistent policy supporting TOD was still lacking at the municipal level.

Competition between municipalities within the region still inhibits TOD, resulting in "no one focus between them," as a transportation planner in Amsterdam noted.

The Netherlands does not take a multidisciplinary approach to TOD: participants in both workshops noted that economists, sociologists, developers, and ecologists are not included in transportation planning. Informal governance bodies, such as TOD committees or advisory groups, are still a foreign concept and interdisciplinary collaboration does not occur (e.g., land use planning, transportation planning, finance, and urban development are all undertaken in separate departments or organizations). However, participants noted that there had been a recent shift toward coordination during the economic crisis, since municipalities could no longer build projects on their own:

> *We're shifting from talking about these issues to actually implementing it, and you also see a shift in the organizations. Different organizations with different responsibilities are talking to each other.* (Transportation planner, Amsterdam)

One example of this was a new bus line approved by Nijmegen, Arnhem, and several smaller municipalities.

The participants expressed a good deal of concern about the role of the national government in TOD. Many indicated that they did not have a good relationship with the national government, which was critical since they finance TOD, outline legal regulations, and develop incentives for local government:

> *It's very unclear to us what they want. They have some kind of implementation plan for public transportation, but nobody knows. They keep it to themselves. So we don't know what it is or how to use it. Because we ask about it and they say well, there's still discussion between the Ministries…they don't want to share.* (Urban/transportation planner, Amsterdam)

Complicating this was the strong history of involvement at the national level, e.g., the National Key Projects, major renovations to the main railway stations that were finalized in policy by the 1990s and were finally being implemented at the time the workshops were held. Participants suggested that the national government had some consistent policy supporting TOD, but they were unstable when it came to implementation (e.g., the 1998 ABC Location Policy that attempted to direct growth to railway station areas). National priority projects are all well and good (e.g., Randstad Urgent (2007) prioritized 33 infrastructure projects in the region), but municipalities also need local transportation infrastructure outside of major railway station areas:

> *…when we [municipalities] have conversations about TOD we almost always focus at the urban nodes or the car-pool level at the most. So it's all about local and regional scale. Everyone at the Ministry is telling me, that's not of national importance. So we leave it to you.* (Urban planner, Amsterdam)

> *For example, the University of Delft is one of the major global, top 20 universities every year. But the national government leaves it to the City of Delft to get a tram line financed. They won't help, whatever.* (Transportation planner, Amsterdam)

This preoccupation with railway stations and unwillingness to take corridor-level initiatives such as BRT or LRT suggest that only some aspects of TOD (e.g., economic development, redevelopment of nationally significant station areas) have been transferred to the national government. Other aspects of TOD (e.g., public health, sustainability) that could be achieved by smaller-scale local planning, such as pleasant walking environments, cycling paths and signals, and mixed-use land development, are not considered essential to TOD in the Netherlands.

Unwillingness to Experiment

At the time the workshops were held, participants felt that there was a lot of stagnation in development that accompanied the prolonged economic crisis. For example, the glut of office space on the market, much of it in transit-accessible areas near railway stations, was not conducive to further development. But one planner in The Hague workshop noted that some municipalities have begun to take risks:

> *In Arnhem-Nijmegen we had some really good city planners, really visionary, and they started a new plan to revitalise old railway stations, 11 of them. So I see even in the smaller towns… now they are starting to see that they can connect these areas, develop on the other side of the station. So if I really saw something in the past seven years it's that we've trained them, they have learned, and now they are proud to do it themselves.* (Transportation planner, The Hague)

Lack of Public Participation

Overall, public participation in the Netherlands is limited to commenting on final plans or projects, which is the most basic level of engagement required by law in many countries. Consultation in earlier phases of the planning process is limited to small-scale, local projects rather than large-scale projects of national significance. One participant in The Hague noted that people would likely get more involved in the local and land use aspects such as walking and cycling paths and well-designed public spaces near stations, rather than the large-scale plans or policies. With a country defined by high-density living and significant rail commuting, Dutch planners may not need to convince community residents of the need for land use and transportation planning to support TOD. One participant noted that if he were to ask people what TOD was, they would have no idea, but they do seem to support new transit infrastructure:

> *…the railway station, they use it. Last year, with the amount of people who took the tram, it's growing by 2.5 to 4 per cent… so TOD as a concept, the implementation and the areas where we do something, they are fond of it.* (Land use/transportation planner, The Hague)

Again, these comments suggest that the land use and design aspects of TOD, common at the smaller and local scales, are not associated with TOD in the Netherlands.

Strengthening the Weaknesses

In the second exercise, participants were asked how they would strengthen key weaknesses in their region in order to overcome barriers to TOD implementation. After a small group exercise, a plenary discussion was held with the entire group. Both the small and large group discussions were digitally recorded, transcribed, and analyzed for relevant themes.

Actor Relationships

Participants in the Amsterdam workshop felt that in order for TOD to be more broadly implemented, a common "story" was needed to align goals between different actors in TOD. One group developed this idea further: the reality of four cities connected by rail in the Randstad region (Amsterdam, Utrecht, Rotterdam, The Hague) could be one element of the story. But the story should also include land use and other issues and could be disseminated to co-workers, professionals outside of planning, and the public.

Involving land use planners, traffic engineers, and others outside of the "usual suspects" in TOD was a common thread. In The Hague workshop, suggestions included anthropologists, sociologists, marketing/communication experts, and lobbyists: these would help TOD experts understand why people would use a certain station or how they meet in public spaces and would also generate new ways of thinking, which was obviously needed to catalyze implementation.

Some discussion of a regional transportation-land use organization came up in both Amsterdam and The Hague:

> *To be more effective you need both informal and formal relationships. But you need the power. Because some other formal actors will only listen to you if you have some form of power, a form of decision-making. That's when they get more serious.* (Transportation planner, Amsterdam)

> *You don't have to formalize to be more effective... I think we need a transport authority on a regional level–that you have to formalize to collaborate with the national government.* (Urban planner, Amsterdam)

However, one participant mentioned that it would likely be ten years before such an organization existed in Amsterdam. In The Hague, a participant expressed the need for municipalities within a region to agree on land goals such as where new office space should go, to avoid situations like the oversupply of office space being experienced at the time the workshops were being held.

Most participants suggested that lower levels of government should have greater responsibilities and the ability to fund transportation improvements, through the ability to generate tax revenue or earmark funds. This would effectively decrease the need for better relationships with the national government, which was a notable weakness in both city-regions.

Public Participation

Although some participants felt that only planners, with their expert knowledge, should make major decisions and develop scenarios to be used in public consultation, several participants mentioned the potential of using games as a way for community residents to understand trade-offs between types of transportation infrastructure or projects. One mentioned the use of mail-in survey cards that had been used to gather responses in a particular neighborhood, but online surveys were not used.

A few participants felt that better marketing of TOD as a concept could get people to see TOD as a means to an end, rather than an end in itself. Another participant mentioned that TOD offered a solution to align different agendas and priorities. Getting different organizations to embrace TOD as a partial solution to problems such as public health or growth management could in fact be more important than spending the time to develop a collective vision for transportation/land use planning in the region. In effect, TOD could be used to help achieve multiple goals, whether it be enhanced transportation, economic growth in municipalities, or better designed public spaces.

Were Policy Ideas/Lessons Transferred?

To understand what the participants learned from the workshop, we used pre- and post-workshop surveys, which asked participants to answer on a five-point scale whether they agreed or disagreed with the statements. Figure 3.4 shows that the workshop resulted in a positive shift from their pre-workshop expectations. More participants indicated that they "will use insights from the workshop in their daily planning practice" in the post-workshop survey than the pre-workshop survey. The exercises helped

Fig. 3.4 Pre- and post-workshop survey results

them understand possible solutions to barriers in TOD implementation and reach a shared vision on the possible solutions. For the statement, "It is difficult to apply policies and ideas from other contexts to the Dutch planning context," responses shifted so that more disagreed with the statement after the workshop. More agreed with the statements that policies and ideas from other planning contexts are useful for identifying weaknesses in transportation and land use planning in their region and would be useful in selecting TOD strategies/options for their region, after the workshop. Most agreed that their organization has the ability to use solutions (policies and ideas) from other planning contexts.

Participants' opinions did not change on some questions: they were just as likely to say they would communicate the results of the workshop to other members of their organizations before and after the workshop. Participants still believed that the culture of the organization where they work does not enable the use of policies and ideas from other planning contexts after the workshop. For the statement, "Conflicting policies between agencies inhibits the use of solutions (policies and ideas) from other planning contexts," post-workshop responses were much more divided.

The participants were also asked to respond to an open question on the most important lesson or insight they took from the workshop. Responses indicated the need for better communication on TOD, which would lead to a broader understanding of the concept, including that "TOD has to come out of the technocratic corner. It is a possibility to address bigger problems in society" and that "We need to broaden the story and simplify at the same time."

While we felt that the policy transfer workshops were a useful way to transfer policy ideas/lessons to practitioners, and the participants seemed to learn ideas from the international cases that they could transfer to their daily practice, it was clear that a number of institutional and cultural barriers to TOD in the Netherlands remain. Dutch planners interested in TOD (the small group of TOD experts) agree that it's a desirable concept, but they're still unsure on how to implement it (e.g., Who is the lead organization? How should the planning process happen? What legal regulations or policies might facilitate TOD?) Here, the lack of willingness to experiment and weak actor relationships are key barriers. TOD is still understood as merely a station area redevelopment concept and not a broader set of policy tools including main street corridors designed to facilitate walking, cycling, and public transit infrastructure such as tram lines and local buses. As the participants themselves noted, this is because only transportation planners with infrastructure development knowledge seem to be aware of TOD. So TOD as a policy concept has not been completely transferred to the Netherlands; policy ideas/lessons on interdisciplinary collaboration, inter-regional collaboration, and integration of land use and transportation planning have not been transferred from other countries. Finally, compared to the US, Canada, and Australia, Dutch planners noted that they had not been trained in public consultation techniques because the practice of planning in the Netherlands is much more top-down than in these other countries. Considering the good level of public support for the higher densities, mixed land uses, and investment in transportation infrastructure necessary for smaller-scale TOD implementation and the national tendency toward dialogue and consensus-building on complex issues such as managing water levels, there is a lot of potential to integrate more public participation into TOD planning.

While Dutch land use and transportation planners are familiar with the TOD concept and the "harder" tools such as land use policies, they're less familiar with the "softer" transferable lessons that are critical in successful implementation, such as good actor relationships, the support of the national government, the need for a multidisciplinary approach, and active public engagement. Participants in the workshops were quite receptive to the "softer" policy ideas from the international cases and intend to bring them to their organizations, which may have some long-term effects on broadening TOD beyond station areas to transit-oriented corridors and regions.

Reflections on the Approach

In our choice of city-regions, we started with a long list of over 30 potential case studies, which were eliminated as they did not meet our three conditions. The most difficult condition to meet was the requirement that three experts in each city-region be able to help us in our evaluation of the cases. Another weakness that we encountered from feedback during the presentation and publication stage of the research was the lack of a single American city-region in our sample. A lot of the innovation in TOD policy, for example on planning tools and strategies to enable TOD and on community benefits to prevent displacement of low-income residents, comes from American cities. Our inability to find local experts stems from our own education and expertise working in Canada and the Netherlands and our research collaboration networks in selected countries.

Using the completed case studies allowed us to compare the 11 cases in just six months, a much shorter amount of time than it would have taken to gather primary data for each case and then analyze it for within-case patterns. The meta-matrix method also allowed the integration of both qualitative (e.g., policy tools, interview quotes) and quantitative (e.g., modal share, population) data. There is no shortage of data to use in meta-analysis, since in-depth case studies are commonly available; the only caveat is that to use a meta-matrix, the cases must be similar. In this study, case study data was coded by a single researcher and discussed with the other researcher on the project. If several researchers were conducting the analysis, codes would need to be developed and agreed upon, which would add time and complexity to the process since developing the codes is an iterative process. Alternatively, a variable-oriented analysis, using a list of variables developed from the literature, could be conducted. This would shape the meta-analysis in a different way and could lead to different results.

Another potential disadvantage to the meta-matrix is that while it was essential in revealing cross-case patterns and distilling the 16 CSFs, it did not tell us which were the most crucial to successful TOD implementation. We attempted to learn this with a rough set analysis, which told us that the most important factors affecting TOD outcomes were national political stability, actor relationships within the region, regional land use-transportation authority, interdisciplinary implementation teams, and public participation (Thomas and Bertolini 2015). Less important factors were policy consistency, key visionaries, and site-specific planning tools. However, we have since found it more useful for the stakeholders in each region to evaluate their own city-region using the full list of 16 CSFs.

We found the policy transfer workshops to be useful, and they could easily be expanded to a wider network of practitioners and academics and to different political, geographic, or social contexts. While it was clear that the participants did not intend to adopt any of the policies, ideas, or institutions completely, they did use the "decontextualized" ideas as inspiration. We did find that cultural and institutional barriers played a role in the policy transfer process: for one thing, transferring TOD ideas to the Netherlands has rarely resulted in concrete actions or outcomes. This may be because close relationships are lacking between the actors involved in TOD, there is little support or direction from the national government, there is little willingness to experiment, and there is no meaningful public participation on TOD because Dutch planners default to basic consultation procedures. Participants also noted that, even though using policy ideas from other places helped them understand the possible barriers to TOD and develop a shared vision on the possible solutions, the cultures of their planning organizations do not favor implementation of these solutions.

Conclusion

This chapter introduced the 11 city-regions that we used to explore cross-case patterns in TOD implementation. The profiles of each case focused on policy consistency, actors and their roles, land use-transportation connections, specific policies and tools, and barriers to TOD. It became clear from the iterative process of developing the coded case reports, profiles, and meta-matrix, that there were in fact persistent cross-case patterns.

These critical success factors told us not only what increased successful implementation of TOD but also what decreased success.

The CSFs then became useful in ranking each city-region ourselves and with the help of three local experts in each case. We then took our findings from the entire set of case studies to Dutch planners, using policy learning workshops to see whether ideas from other case studies could jump-start creative thinking on TOD. We found that local planners were more receptive to some ideas than others, due to their limited definition of TOD, lack of engagement with the public, weak actor relationships, and unwillingness to experiment with new administrative arrangements.

While there are many things that could have turned out differently had we used different methods in our case study comparison and analysis, we believe that our findings are still robust enough to be shared with other land use and transportation planners. In particular, our list of critical success factors could be used to generate discussions on policies and the development of stronger relationships among the stakeholders responsible for TOD implementation. Barriers to policy learning exist, including differences in cultural, political, geographic, and legal contexts. Dutch planners, as we have learned, have a fairly limited understanding of TOD as an infrastructure concept, as opposed to a broader approach to land use-transportation integration, public health, and growth management. Planners may be trained differently, resulting in different skill sets related to TOD: in the Netherlands, less training in public engagement has resulted in more basic consultation processes.

In the next chapter, we discuss persistent barriers to the implementation of TOD and potential solutions.

References

Adenot, F. J. (2007a). *Ville de Longueil: Place Charles-Le Moyne*. Recherche URBATOD etude de cas, Departement d'etudes urbaines et touristiques, Universite de Quebec a Montreal (UQAM).

Adenot, F. J. (2007b). *Ville Mont-Royal: Les entreprises ferroviaires orientant le developpement urbain*. Recherche URBATOD etude de cas, Departement d'etudes urbaines et touristiques, Universite de Quebec a Montreal (UQAM).

Allum, P. (2003). The politics of town planning in post-war Naples. *Journal of Modern Italian Studies, 8*(4), 500–527.

Bertolini, L. (1998). Station area redevelopment in five European countries: An international perspective on a complex planning challenge. *International Planning Studies, 3*(2), 163–184.

Bertolini, L., & Spit, T. (1998). *Cities on rails: The redevelopment of railway station areas.* London/New York: Spon/Routledge.

Bertolini, L., Curtis, C., & Renne, J. (2012). Station area projects in Europe and beyond: Towards transit oriented development? *Built Environment, 38*(1), 31–50.

Bull, A. C. (2005). Democratic renewal, urban planning and civil society: The regeneration of Bagnoli, Naples. *South European Society and Politics, 10*(3), 391–410.

Cascetta, E., & Pagliara, F. (2008). Integrated railways-based policies: The Regional Metro System (RMS) project of Naples and Campania. *Transport Policy, 15*(2), 81–93.

Cervero, R. (1998). *The transit metropolis: A global inquiry.* Washington, DC: Island Press.

Chapple, K., & Loukaitou-Sideris, A. (2019). *Transit-oriented displacement or community dividends?* Cambridge, MA/London, UK: MIT Press.

Chorus, P. (2012). *Station area developments in Tokyo and what the Randstad can learn from it.* Amsterdam: University of Amsterdam.

Chorus, P., & Bertolini, L. (2016). Developing transit-oriented corridors, insights from Tokyo. *International Journal of Sustainable Transportation, 10*(2), 86–95.

Curtis, C. (2008). Evolution of the transit-oriented development model for low-density cities: A case study of Perth's new railway corridor. *Planning Practice and Research, 23*(3), 285–302.

Curtis, C., & Scheurer, J. (2012). *Benchmarking public transport accessibility in Australian cities.* Paper presented at the Australasian transport research forum.

Curtis, C., Renne, J. L., & Bertolini, L. (Eds.). (2009). *Transit oriented development: Making it happen.* Farnham/Burlington: Ashgate Publishing.

Dolowitz, D., & Marsh, D. (2000). Learning from abroad: The role of policy transfer in contemporary policy making. *Governance, 13*(1), 5–24.

Duffhues, J. (2010). *Transportation as a means for densification–Or the other way around?* Paper presented at Colloquium Vervoersplanologisch Speurwerk, Roermond, The Netherlands.

El-Geneidy, A., Kastelberger, L., & Abdelhamid, H. T. (2011). Montreal's roots: Exploring the growth of Montreal's indoor city. *Journal of Transport and Land Use, 4*(2), 33–46.

Filion, P. (2001). Suburban mixed-use centres and urban dispersion: What difference do they make? *Environment and Planning A, 33*(1), 141–160.

Filion, P., & McSpurren, K. (2007). Smart growth and development reality: The difficult co-ordination of land use and transport objectives. *Urban Studies, 44*(3), 501–523.

Filion, P., McSpurren, K., & Appleby, B. (2006). Wasted density? The impact of Toronto's residential-density distribution policies on public-transit use and walking. *Environment and Planning A, 38*(7), 1367–1392.
Gelli, F. (2001). Planning systems in Italy within the context of new processes of 'regionalization'. *International Planning Studies, 6*(2), 183–197.
Grant, J. (2002). Mixed use in theory and practice: Canadian experience with implementing a planning principle. *Journal of the American Planning Association, 68*(1), 71–84.
Grin, J. (2010). The governance of transitions. In J. Grin, J. Rotmans, & J. Schot (Eds.), *Transitions to sustainable development: New directions in the study of long term transformative change* (pp. 265–285). New York/London: Routledge.
Hartoft-Nielsen, P. (2007). Deconcentration of workplaces in greater Copenhagen–successes and failures of location strategies in regional planning. In E. Razin, M. Dijst, & C. Vaquez (Eds.), *Employment deconcentration in European cities (springer GeoJournal library 91)* (pp. 53–87). Dordrecht: Springer.
Keurs, K., Maat, K., Rietveld, P., & De Visser, G. (2012). *Transit oriented development in the Randstad south wing: Goals, issues and research.* Paper presented at the Building the Urban Future and Transit Oriented Development Conference, Paris.
Knowles, R. D. (2012). Transit oriented development in Copenhagen, Denmark: From the finger plan to Ørestad. *Journal of Transport Geography, 22*, 251–261.
Langendijk, A., & Boertjes, S. (2012). Light rail: All change please! A post-structural perspective on the global mushrooming of a transport concept. *Planning Theory, 12*(3), 290–310.
Legacy, C., Curtis, C., & Sturup, S. (2012). Is there a good governance model for the delivery of contemporary transport policy and practice? An examination of Melbourne and Perth. *Transport Policy, 19*(1), 8–16.
Meijers, E., Hollander, K., & Hoogerbrugge, M. (2012). *Case study metropolitan region: Rotterdam-The Hague.* The Hague: European Metropolitan Network Institute/European Regional Development Fund.
Mouritz, M., & Ainsworth, L. (2009). Successful delivery mechanisms: Coordinating plans, players and action. In C. Curtis, J. L. Renne, & L. Bertolini (Eds.), *Transit oriented development: Making it happen* (pp. 171–183). Burlington: Ashgate Publishing.
Naess, P. (2005). Residential location affects travel behavior—But how and why? The case of Copenhagen metropolitan area. *Progress in Planning, 63*(2), 167–257.
Naess, P., Strand, A., Naess, T., & Nicolaisen, M. (2011). On their road to sustainability? The challenge of sustainable mobility in urban planning and development in two Scandinavian capital regions. *Town Planning Review, 82*(3), 285–316.

Newman, P. (2007). Planning for TOD in Australian cities. *Building Environment Design Professionals Environment Design Guide, 2*(15), 1–11.

Nijkamp, P., van der Burch, M., & Vindigni, G. (2002). A comparative institutional evaluation of public-private partnerships in Dutch urban land-use and revitalisation projects. *Urban Studies, 39*(10), 1865–1880.

Pagliara, F., & Papa, E. (2011). Urban rail systems investments: An analysis of the impacts on property values and residents' location. *Journal of Transport Geography, 19*(2), 200–211.

Perl, A., & Pucher, J. (1995). Transit in trouble? The policy challenge posed by Canada's changing urban mobility. *Canadian Public Policy, 21*(3), 261–283.

Programmabureau StedenbaanPlus. (2011). In P. StedebaanPlus (Ed.), *StedenbaanPlus.* Den Haag: Bestuurlijke Platform Zuidvleugel.

Raad, T., & Kenworthy, J. (1998). The US and us. *Alternatives Journal, 24*(1), 14–22.

Schwanen, T., Dijst, M., & Dieleman, F. M. (2004). Policies for urban form and their impact on travel: The Netherlands experience. *Urban Studies, 41*(3), 579–603.

Spaans, M., & Louw, E. (2009). *Crossing borders with planners and developers and the limits of lesson-drawing. City futures in a globalising world.* Madrid: University Rey Juan Carlos of Madrid.

Tan, W. G. Z. (2009). *Policy context of key transit-oriented projects for station locations within the Netherlands.* Paper presented at TU Berlin, Centre for Metropolitan Studies.

Tan, W. (2011). *NICIS KEI case study #1: Perth, Western Australia.*

Tan, W., Bertolini, L., & Janssen-Jansen, L. (2012). *The role of implementation barriers and institutional incentives in transit-oriented development: The cases of Perth, Portland and Vancouver.* Paper presented at the AESOP 26th annual congress. Ankara, Turkey.

Thomas, R., & Bertolini, L. (2015). Policy transfer among planners in transit-oriented development. *Town Planning Review, 86*(5), 537–560. https://doi.org/10.3828/tpr.2015.32.

TransLink. (2012). *Transit-oriented communities guidelines. Creating more livable places around transit in Metro Vancouver.* Vancouver: TransLink.

Van Egmond, P., Nijkamp, P., & Vindigni, G. (2003). A comparative analysis of the performance of urban public transport systems in Europe. *International Social Science Journal, 55*(176), 235–247.

CHAPTER 4

Persistent Challenges and Potential Solutions: Equitable TOD

Abstract It can be valuable to learn about policy ideas from other cities and regions, but barriers to implementation may still exist. Persistent barriers to TOD include the loss of affordable housing, cost, zoning and other regulatory issues, policy consistency and planning coordination, and public opposition. In this chapter, we give examples of challenges and solutions from municipalities other than our 11 case studies, broadening the range of potential policy ideas and concepts that can be used as learning and inspiration. In particular, we focus on the challenge of making TOD equitable, which has been a challenge in many countries in recent years and a key area of policy innovation in the US. These solutions can help address some of the persistent challenges to TOD implementation that we observed both in the international literature and in our workshops with Dutch stakeholders.

Keywords Equity • Transportation infrastructure • Affordable housing • Displacement • Policy tools

Transit-oriented development is, itself, a policy concept that has become internationally known, emulated, and copied. As we have seen, it can be valuable to learn from other cities and regions, but barriers to implementation may still exist including the culture of organizations, the planning practice in a particular country, the nature of the relationships between institutional actors, or other factors unique to the local context.

R. Thomas, L. Bertolini, *Transit-Oriented Development*,
https://doi.org/10.1007/978-3-030-48470-5_4

Now that we have presented the results from our 11 case studies and shown how others can learn from them in practice, we would like to reflect on other factors that may play a role in TOD implementation. At the beginning of this book, we discussed a number of barriers to TOD: the loss of affordable housing, cost, zoning and other regulatory issues, policy consistency and planning coordination, and public opposition. Here we continue this discussion, reiterating concerns about TOD which did not emerge in the meta-analysis, potentially because of our choice of cases. This allows us to give examples of challenges and solutions from municipalities other than our 11 case studies, broadening the suite of potential policy ideas and concepts that can be used as learning and inspiration in other cities and regions. In particular, we focus on the challenge of making TOD equitable, which has been a challenge in many countries in recent years and a key area of policy innovation in the US.

It is worth noting that a lot of the literature cited in this chapter is from the US, for several reasons. First, the US is a very large country with an extensive network of fixed-route transit (heavy rail, light rail, sometimes also including BRT); therefore, there are abundant examples of TOD implementation in the US. Second, the dominance of the American examples of TOD in the literature is undoubtedly related to our reliance upon English language articles. Our inclusion of American solutions here potentially offsets the lack of American cases in our study sample. Third, the fact that the private market supplies most housing in the US means that housing affordability is a persistent issue, particularly in large metropolitan areas and those served by rail. As a result, specific tools and strategies have been introduced to encourage the construction of affordable housing in key areas, including areas that are well-served by transit. Fourth, racism and segregation in American cities have resulted in pervasive disinvestment in low-income neighborhoods. When new infrastructure is proposed and built in these areas, it inevitably causes community change and, in some cases, displacement through gentrification; Chapple and Loukaitou-Sideris (2019) refer to this as "the dark side of TOD." And finally, the planning process and culture in the US includes extensive public participation: activist- and community-led protests around transportation date back to the 1950s, and in recent years community members and organizations have been instrumental in reframing transportation projects, including TODs.

So, although many of the solutions presented in the rest of the chapter are obviously rooted in the political, geographic, and planning context of the US, we present them here because they help address some of the persistent challenges to TOD implementation that we observed both in the international literature and in our workshops with Dutch stakeholders.

Persistent Equity Challenges

Loss of Affordable Housing in Station Areas and Corridors

One of the major benefits of public transit is that it connects people to work, education, health care, and other services offered in metropolitan regions, which is particularly critical for people who are low-income and do not own cars. However, transportation authorities and municipal transportation planning departments have not always prioritized the needs of these transit users, which include seniors, youth and young adults, immigrants, and visible minorities. As briefly mentioned in Chap. 1, transportation infrastructure projects have a history of negative impacts on communities, including ignoring community values in planning processes, division and separation of communities through large-scale infrastructure, and massive displacement of inner city residents during the 1950s and 1960s. There is quite an extensive literature on this topic, particularly in the field of planning, where environmental justice and equity have been significant concerns for decades.

In recent years, the increase in construction and extension of mass transit systems in areas where they did not exist before has had major impacts (e.g., in American, Canadian, and Australian cities without extensive rail-based and other transit infrastructure) on communities. For example, in Denver, a transit expansion program will add 122 miles of rail, 18 miles (29 km) of BRT, and 55 transit stations (Bardaka and Hersey 2018); 29 percent of Metro Denver apartments proposed or under construction in 2017 were located within a half-mile (800 m) of a rail station. This type of focused growth can have positive impacts on regions that would like to slow the pace of suburban and exurban sprawl.

While all of this construction is great news for cities and regions that seek sustainable alternatives to single-occupant vehicle travel, TOD also has a number of negative impacts. Investment in TOD might force authorities to invest less in bus routes since capital and operational funding are limited (Lung-Amam et al. 2014) and the development process can be disruptive to local businesses, who face declining sales during construction and may face higher rents once construction is complete. Many communities are concerned that the demand for housing near transportation infrastructure could cause low-income residents to be displaced. And since transit authorities in the US rely on low-income users, who take transit at a much higher rate than higher-income users, "Transit agencies may then find themselves the victims of their own expansion, setting in motion a speculative real estate market that delivers high-rent land uses but few new transit riders" (Bardaka and Hersey 2018). As Chapple and Loukaitou-Sideris (2019, 40) point out, displacement has deep roots in the United States, "as well as other countries that have experienced colonial aggression, and subsequently land theft and exclusion, against indigenous and/or minority populations." They note that in countries such as Cambodia, the Philippines, Ethiopia, India, and Brazil, railway projects and their related developments led to the forced eviction of residents and in some cases their relocation to the periphery of regions.

In Canada, Grube-Cavers and Patterson (2015) found a significant relationship between gentrifying areas and the proximity to rapid transit and adjacent gentrifying areas in Vancouver and Toronto, and Jones (2015) found a loss of affordable housing stock along the SkyTrain corridor in Burnaby due to policies that promote TOD. Espeseth's (2017) study of a proposed TOD node (Joseph Howe Future Growth Node) in Halifax showed a high proportion of publicly owned land and non-profit housing, which could make it vulnerable to a loss of affordable housing as the transit infrastructure is completed. To prevent this, the city needs to improve coordination between planning for affordable housing and planning for transit, including land acquisition, accompanied by stable funding to develop or renovate housing.

In a study of over 2400 mixed-income households in multifamily apartments within a 10-minute walk of rail stations in Denver, people who lived in market-rate housing were much more likely to use a vehicle for their most frequent travel mode: 69 percent did so, compared to just 34 percent of residents of low-income households (Bardaka and Hersey 2018). For those who had an employer-provided transit pass, just 28 percent of market-rate residents commuted by transit while 92 percent of the low-income residents did. In a comparison of 345 station areas in the Boston area (including stops on high-frequency bus routes), the station areas that performed best had a combination of built environment, social, and transit attributes that reduce driving, increase transit ridership, and promote equity (Pollack et al. 2014). In fact, four measures explained about 87 percent of the difference in household driving: the proportion of the residents who do not own cars, percentage of households with incomes below $25,000, percentage of renter-occupied units in the station area, and affordability based on the Center for Neighbourhood Technology's Housing + Transportation Index.

In a comparison of 4400 fixed-route transit stations across 39 regions in the US, Renne et al. (2016) classified each as a TOD (those with a Walk Score of 70 or higher and a gross housing density of at least 8 units per acre), a hybrid (those that meet one of the criteria), or a transit-adjacent development (those that do not meet either criterion). They found that TODs are more expensive places to own or rent, but are more affordable than hybrids or transit-adjacent developments because of the lower cost of transportation to households (e.g., transit use rather than car ownership). TOD station areas, station areas with higher levels of transit frequency, stations closer to the city center, and station areas with more people and higher job densities were frequently associated with lower levels of household transportation spending. The authors believe there is enormous potential to build out existing transit station areas, since currently only 36 percent of station areas had densities over 8 units per acre, and transportation costs should remain low as a percentage of household income if residents use transit. However:

> *…whereas transportation costs in TOD are affordable, the housing market in TOD is quickly escalating. Unless policies are implemented soon to bolster densities in underutilized fixed-route transit station areas, the limited amount of housing in TOD is likely to continue to be priced out of the reach of average Americans.* (Renne et al. 2016, 832)

Impacts on Local Businesses

Chapple and Loukaitou-Sideris (2019) also identified four types of commercial gentrification in international literature: retail upscaling, tourism gentrification, art districts, and transit-oriented districts. Retail upscaling involves a new type of customer being attracted to the area, and the new shops and services are often at odds with the historic or community character (e.g., shops providing higher-end organic or gourmet foods). This can have the effect of excluding long-term residents and business owners. Tourism gentrification occurs when local business associations, municipalities, or private-sector developers seek to benefit from the area's ethnic cultures for their own benefit, which can paradoxically increase rents in the area and displace the ethnic population. The special designation and marketing of arts districts has ended up displacing local artists from local areas. In TOD, San Francisco's Mission District saw an increase in shops catering to regional rather than local needs and a higher rate of closure among Hispanic-owned businesses.

In Los Angeles, opening of the Chinatown transit station in 2003 has had long-term effects on the small-scale local businesses owned and operated by residents (Chapple and Loukaitou-Sideris 2019). Interviews revealed that community activists and shop owners were concerned about the closure of several long-standing anchor businesses and opening of new businesses that cater to new, wealthier customers. In San Francisco, the Temescal neighborhood saw a 15 percent decrease in ethnic businesses, an increase in strip mall retail, and more restaurants and cafés. Interviews showed that while rising rents were a factor, the loss of local clientele was another: as residents were displaced, the shops could no longer rely on local buyers.

Equity Concerns = Equitable Solutions

While these issues may be products of the American political, cultural, and geographic context, they are worth considering because similar mechanisms may be at play in other countries, and more equitable TOD is more complete TOD. The City of Milwaukee, in a recent TOD area plan (2018, 11), noted that:

> *There is also a growing realization that equity needs to be at the foundation of planning for Transit Oriented Development. While new development has many positives, including new housing and shopping options in the community and increased tax base for the City, the benefits of new development—including TOD—are not always distributed equally…At the most fundamental level, equitable growth means that development benefits and does not displace either current residents or the cultural character of neighborhoods and that historically disadvantaged groups are able to gain access to wealth building opportunities created by investments in transit and Transit Oriented Development.*

A more holistic approach to TOD is, frankly, lacking in many cities and countries around the world. Many American cities and regions advocate equitable TOD, or eTOD for short. eTOD "offers one solution to this investment-revenue disconnect by integrating income-restricted housing into station-area developments that are otherwise likely to attract market-rate housing" (Bardaka and Hersey 2018). In the following section, we present a number of tools, strategies, planning regulations, and processes that have led to more equitable TOD in American cities and regions and could be sources of reflection or inspiration for others. This focus on equitable distribution of the benefits derived from TOD, as well as the negative aspects, was just beginning to take shape when we conducted our study, described in the previous chapter, and was not an acknowledged factor in successful implementation within the case studies we studied. eTOD considerations have now become fairly well integrated into American TOD approaches.

The surprising thing is that many of the tools and strategies used to promote more equitable TOD also address many of those persistent barriers to TOD that we uncovered in the literature: besides protecting affordable housing, they address impacts on local businesses, the costs of TOD, planning regulations, public opposition, and support from higher levels of government.

Tools and Strategies to Enable More Equitable TOD

Evaluating the Impact of Transportation Investments

In the US, there are several tools to help planners, developers, and community organizations understand the potential impact of transit investments on specific neighborhoods. The Center for Neighborhood Technology's Housing + Transportation Index, which was launched in 2006, allows users to see areas that have low transportation and low housing costs: areas with good transit access where housing is still affordable. The index is now available for 917 metro- and micropolitan areas covering 94 percent of the US population (Mueller et al. 2018). The US Department of Housing and Urban Development has a Location Affordability Index and Portal, which illustrate the costs of housing and transportation as a percentage of family income for different family profiles. The Dukakis Center for Urban and Regional Policy developed a rating system for equitable TOD: eTOD scores assess the transit supportiveness of a station area focusing on transit service and accessibility, orientation of the nearby population toward transit, and the character of the built environment. Several regions, such as Denver, have developed regional equity atlases to illustrate transit accessibility to job or service destinations.

Mueller et al.'s (2018) corridor housing preservation tool compares and prioritizes transit and commercial corridors as sites for preservation of existing affordable rental housing. The authors identified three types of corridors:

- Corridors where substantial stock of affordable units, great access to jobs, and relatively low development pressure mean that acquiring and preserving buildings could be cost-effective
- Corridors where relatively high development pressure, good job access, and dwindling stock of affordable units suggest the need for urgent action to prevent total loss of affordability
- Corridors where low wage job access via transit is relatively poor so preservation is questionable

These tools can help municipal and state governments understand where residents or local businesses are likely to be displaced as a result of new transportation infrastructure and where to focus their efforts in preserving or building new affordable housing. In order to be adapted to international contexts, these tools would need to be rebuilt using national databases (e.g., census data, national travel data) from the relevant country.

Buying/Holding Properties Near Transit Infrastructure

To protect affordable housing or enable the construction of new units, many cities and regions have decided to invest in buying sites located near future transit infrastructure, which can be developed later by either non-profit organizations or developers interested in building affordable housing. This is particularly relevant in the US, Canada, Australia, the UK, and other countries where most housing is supplied by the private sector. The situation in European countries, with much higher rates of publicly supplied housing, is quite different. However, these strategies often incorporate a regional approach to urban development that is critical to TOD, which could be combined with efforts to preserve and build new affordable housing and growth management strategies designed to focus urban growth into areas where high-quality infrastructure and services already exist. For example, the Urban Land Conservancy in Denver acquired properties in areas where transit infrastructure was scheduled to be built and held the properties until station area rents rose enough to attract affordable housing developers.

American states can issue tax credits for buying, rehabilitating, or building new rental housing for low-income households. While the units do not necessarily have to be located near transit infrastructure or routes, about half of the state housing finance agencies provided additional points in their scoring criteria for projects near transit and half provided points for access to neighborhood amenities and resources. However, the Poverty and Race Research Action Council (2015) reported that in the past 20 years, only 16 percent of affordable developments funded through low-income tax credits are located within a half-mile (800 m) of a transit station and 4.3 percent within a quarter-mile (400 m). The high cost of land and NIMBY (not in my backyard) attitudes to affordable housing were the main barriers.

When Portland built its Yellow Line, which was supposed to serve lower-income and minority neighborhoods, transportation authority TriMet used surplus Federal Transportation Administration money to buy properties in the neighborhood (the line was completed under budget). Portland Development Commission and city land use planners advocated for the final alignment and station locations, worked with TriMet to identify and procure potential development sites along the corridor, and worked with community members to reassure them and address concerns about high-rises and gentrification (Poverty and Race Research Action Council 2015). TriMet's real estate team bought properties, rights-of-way for stations and railways, and construction staging areas, many of which were vacant, underused, or blighted properties. TriMet's property management/development policy focuses on enhancing ridership and increasing housing availability and services for low- and moderate-income households. Its supplementary TOD policy requires that developments on TriMet property maximize density, reduce car dependency, activate public spaces, and support community aspirations. Patton Park, a 54-unit affordable housing project built on one of the sites, offers units for households with 50 percent of the area median income for all units and publicly subsidized three-bedroom units for families under 30 percent of the area median income.

In Seattle, regional authority Sound Transit deeply discounts property sales to make affordable housing projects viable. A 2015 statute in Washington stipulated that Sound Transit had to begin considering affordable housing outcomes in its capital projects: 80 percent of its surplus property must first be offered to a non-profit development, local jurisdiction, or housing authority. The bidding entity must agree to supplying over 80 percent of the units for people making 80 percent of area median income or less (TransitCenter 2018). Although Sound Transit owns a lot of properties, most of the projects so far have been in King County, not Pierce and Snohomish Counties, which may not financially be able to fill in gap financing for projects until developers are ready to begin construction.

Loans/Programs Addressing the Loss of Affordable Housing and Local Businesses

Cost is a persistent barrier to TOD, including the high cost of land and the complexity of planning in built-up areas rather than on greenfield sites. Many cities have created loans and other funding programs that encourage TOD while equitably distributing benefits. On the housing side, since it is more difficult to build affordable housing than market-rate units, developers usually opt to build market-rate projects. Cities and regions have stepped in to make affordable housing development more financially viable for developers with grants and loans.

Enterprise Community Loan Fund partnered with public- and private-sector investors to create the Denver Regional TOD Fund in 2010 to create and preserve affordable housing in future transit corridors in the city. Funds are available to borrowers in seven Metro Denver counties to acquire property for affordable housing and supportive commercial space. One investor in the fund is Colorado Housing Finance Authority, which tailored its low-income housing tax credit program to benefit eTOD projects. Other investors include non-profit organizations, several state departments including housing and economic development, and two major US banks. The City and County of Denver contributed resources to create or preserve affordable housing in station areas (Bardaka and Hersey 2018). The fund has already granted 16 loans and 11 have been repaid, resulting in 1324 affordable homes, a new public library, and over 100,000 sq. ft. (9290 sq. m) of supportive commercial and non-profit space that have been created near transit (Enterprise Community 2018).

Metro Los Angeles' MATCH Loan Program provides loans to developers at a low interest rate to build affordable housing within a half-mile (800 m) of transit lines. The funding allows developers to develop vacant land or buy existing housing to keep it affordable for people making 60 percent of the area median income. Metro LA does not own much land, so they opted to provide financing instead. Their contribution of $9 million was matched by the California Endowment and Local Community Development Financial Institutions, bringing the total funding to $75 million, enough to create or preserve 1500 units.

In Indiana, the Indianapolis Neighborhood Housing Partnership (INHP), Cinnaire, and JPMorgan Chase partnered with the City of Indianapolis and other financial institutions to launch the city's first equitable transit-oriented development fund in 2019 (Carlstedt and Washburn 2019). The $15 million fund is a deliberate effort to preserve or create 1000 housing units within a half-mile of transit stops to create a variety of affordable housing options within specific growth areas. INHP will borrow from the fund to buy sites to use for mixed-use mixed-income housing: they determined that there are seven regionally significant walkable urban neighborhoods within the proposed station areas of the new Red Line, and four of them already have market rents that are too high for low- and middle-income families. JPMorgan granted $500,000 to the eTOD fund as part of their PRO Neighborhoods program, a five-year community development program. Cinnaire, a community development finance partner that develops partnerships with mission-focused organizations and investors to achieve community-driven revitalization, manages the fund.

To prevent commercial displacement, Seattle's Community Development Fund was created four years before construction began on its MLK line. Funding and technical assistance were provided to help businesses stay open, including help with marketing, access plans, signage, façade improvements, bookkeeping, and legal issues (Chapple and Loukaitou-Sideris 2019). Payments were made to businesses that were forced to relocate or had their operations disrupted due to construction. In Minneapolis and St. Paul, the Ready for Rail Business Support Fund provided forgivable loans to small businesses that could show a decrease in sales due to light rail construction in the Central Corridor. Loans were also created for off-street parking, marketing and buying campaigns, and other technical assistance. A partnership of businesses, non-profit community developers, and local and regional governments was also developed to support business and property owners in the corridor: almost five years after construction, almost two-thirds of businesses found that employment and wage levels had not changed, though half reported some decreases in the number of customers.

Planning Regulations Encouraging Equitable Development

Zoning and regulatory issues have long been barriers to TOD, as well as lack of planning coordination. In many cities, the type of dense, mixed-use, multifamily development critical to successful and equitable TOD is simply not allowed. Local authorities are using a number of tools to ensure that TOD is encouraged and that they are fair and beneficial to local residents and businesses.

The City of Denver reduced parking requirements in station areas and promotes complementary TDM principles in project development. For example, in Denver at 38th and Blake Station, ULC leveraged a $1.53 million TOD fund loan to acquire a vacant property and sell it to an affordable housing developer, whose project received a Colorado Housing Finance Authority low-income housing tax credit award to support households earning 60 percent area median income or less (Bardaka and Hersey 2018). The City and County of Denver adopted a citywide development fee and introduced up-zoning to the station area, encouraging market rate developers to build mixed-income housing in return for greater density. This allowed another market rate developer to contribute to the project, which allowed the affordable project to provide units for households earning 30 percent or less of the area median income.

The City of Austin passed a TOD ordinance in 2005, designating the boundaries of future Red Line station areas as TOD zones and establishing an interim overlay zone to ensure developments would be supportive of transit and pedestrian environments. It also set guidelines (goals, not requirements) for the preparation of station area plans that would include a housing affordability analysis and description of strategies to provide at least 25 percent of new rental housing to be affordable to households at or below 60 percent of median family income. The city allowed density bonuses for developers that met affordable housing goals. The M Station development site had a minimum density of 2 stories and 45 units per acre, which could be waived in exchange for providing affordable housing. The project was funded by a low-income tax credit and a loan from the municipal housing finance authority to provide housing at 30 percent of the median household income. The City also created a program that provides fee waivers and expedited review of development applications for projects that meet the criteria of being safe, mixed-income, accessible, reasonably priced, transit-oriented, and meeting green building standards.

Technical assistance from the US Federal Transportation Administration helps cities build "equitable, compact, mixed-used development around transit stations or along transit corridors, with a focus on development in economically disadvantaged areas" (Federal Transportation Administration 2017). Omaha, Nebraska, used this assistance to integrate TOD policy language into their plans: several of their area plans for BRT stations recommend that new and infill development should add to the affordable housing stock, at a minimum replacing these units at a 1:1 ratio. Units should be at a price point reasonable to households making the area's median income (City of Omaha 2019a, b, c).

Charlotte, North Carolina, used FTA assistance to pass new TOD regulations in 2019 to encourage new construction in their existing TOD districts. A new density bonus will allow developers to add height if at least 10 percent of the units are affordable on each floor above the maximum height. Or they can pay into the City of Charlotte Housing Trust Fund (Chemtob 2019). There are now four TOD districts; e.g., the Transit Urban Center district allows heights up to 130 feet (10 stories), but developers can build up to 300 feet (23 stories) under the bonus height system, unless the building is within a quarter-mile of a rapid transit station, in which case the height is unlimited. There are 770 acres zoned for TOD along LRT lines, and the city was rezoning properties to fit the new regulations in 2019.

The City of Milwaukee neighborhood plan for Walker's Point (2018), a neighborhood situated on a proposed streetcar line extension, specifies the number of new units which will be built in the neighborhood, the proposed number of new storefronts and new jobs. Maximum and minimum building heights around the proposed stops will be increased in the zoning code (e.g., increasing building heights up to 13 stories in existing commercial and industrial mixed zones, revised minimum lot area per dwelling unit to allow more density and smaller units), and new zoning districts will allow live-work and home occupations and duplexes, triplexes, and fourplexes in the existing residential zones. The implementation strategy includes the development of a land banking strategy or land acquisition fund, which would allow the City to buy sites and then contribute or sell them at a discount to projects that meet defined goals. The City has proposed new guidelines for the use of tax increment financing to support affordable and mixed-income housing within TOD.

Chicago's TOD ordinance was enacted in 2013 and successfully led to development near stations. The City passed an amendment to the municipal code governing TODs in January 2019. It modifies transit-served locations to include projects near more than 20 CTA-designated high-ridership and high-frequency bus routes along eight major corridors and corridor segments. It also requires the City to develop an Equitable TOD Implementation Policy Plan by 2020. The goal of the amendment was to extend the benefits to residents and businesses without rail transit. The eTOD Implementation Policy Plan "will include strategies aimed at avoiding displacement of residents and businesses, investing in low-income communities and communities of color, supporting transit investment, and ensuring appropriate density levels and parking requirements to neighbourhood context"(Grimshaw 2019).

Addressing Public Opposition: Building a Collaborative Practice

Public opposition is a persistent challenge in TOD, and our Dutch workshops showed that basic public engagement and poor actor relationships were considered weaknesses. Several cities and neighborhoods have been able to work with community organizations, activists, and different levels of government in the creation of TODs:

> Equitable TOD is a process *that requires residents to have the opportunity to come up with their own visions of their desired communities, and be heard. It is a collaborative, and often messy, process of coming to terms with the wants and needs of communities, and ensuring that development helps them to achieve their goals.* (Lung-Amam et al. 2014, 3, italics original)

Sandoval (2016) describes how community activists and organizations in Logan Heights (San Diego) and Boyle Heights (Los Angeles) fought TOD proposals and ultimately became full partners in the development of the projects. In Boyle Heights, the displacement of residents through a new mixed-income housing project, which replaced an existing public housing project, resulted in the mobilization of the community. Community organizations pushed the housing authority to provide adequate displacement resources and forced them to add a clause stating that residents would get to move into the development, inspect it when they moved in, and have a week to change their minds if it didn't work for them, rather than having to accept whatever unit the housing authority gave them. One community organization, the East Los Angeles Community Corporation, redeveloped a historic hotel into 50 units of affordable housing. In Logan Heights, neighborhood resistance to the gentrification and neighborhood turnover represented in the El Mercado Del Barrio project resulted in 93 affordable units, murals built into the project, a plaza, and a supermarket catering to the Latino market. A new gateway, designed by an esteemed local artist, welcomes people to the neighborhood, and historic murals are now protected under National Historical Preservation mandates and cannot be moved or destroyed.

In Fruitvale (Oakland), local community organization Unity Council acted as a catalyst for neighborhood transformation (Sandoval 2018). They led the development of the Fruitvale Transit Village TOD, including a strong desire to protect Latino retail shops. Strong connections to all three levels of government enabled them to secure funds to develop the project, which is now a model of eTOD in low-income communities. Unity Council initiated community outreach, working with city officials, Bay Area Rapid Transit, the housing authority, politicians, and local planners. In all three neighborhoods, activist organizations had a long history of neighborhood action and community work that engendered trust and legitimacy with residents, as well as informal political networks with urban planning agencies, philanthropy, and government. They were able to establish concrete demands, advocate for specific planning interventions, and rely on progressive local politicians. Ethnic identity materialized into a political agenda of redistribution: increased affordable housing, access to more social services, improved public spaces, and links to regional transit. In Sandoval's words:

> *Planners need to listen and work alongside low-income communities and help them define what equity means to them in that particular time and space. Equity comes out of context, particular history, sense of place, and specific forms of oppression. Hence, in each neighborhood, making amends for historical oppression will materialize in different culturally symbolic spatial meanings.* (2018, 12)

To pass Chicago's recent TOD ordinance amendment, members of Elevated Chicago and its Systems Change Working Group began engaging the Mayor's Office, City Departments of Transportation and Planning and Development, and Chicago Transit Authority. Recommendations came from the Systems Change Working Group "and were shaped by the priorities of its transit-served communities supporting equitable TOD" (Grimshaw 2019). The group is co-chaired by the Center for

Neighbourhood Technology, which also works directly with communities to promote eTOD and maintains an eTOD web portal. This helps developers, community organizations, and other stakeholders fully understand the social and economic benefits of development near transit options. Bus ridership in Chicago has declined recently, and CTA believed that more incentives to develop and live near bus routes could boost ridership: more residents live near bus routes than rail stations. Chicago is the only American city to propose TOD along high-frequency, high-ridership bus corridors.

In Milwaukee, the City developed a TOD plan for two neighborhoods, Bronzeville and Walker's Point, where new streetcar stops will be built in the future. Funded by a $750,000 federal grant, the study that led to the development of the plan included ten community workshops, five plan advisory group meetings, and information sessions at community events to collect feedback from 1800 people. Bilingual translators were available and materials were produced in Spanish. The City tried to ensure the staff working on the study represented the races in the community, and they wanted minority-led firms working on the project. The City also partnered with two community organizations, who advertised meeting times and locations and helped with interviews with local leaders and organizations. The City is seeking federal funds to support the expansion of the streetcar system into both neighborhoods, but:

> *Even if the streetcar never materializes, the plans are intended to create more walkable and transit-friendly places…both plans include changes to east-west running streets that intersect with the streetcar corridors to improve bicycle and pedestrian friendliness.* (Jannene 2018)

The new TOD plan (City of Milwaukee 2018) reinforces the equity priorities of several area plans, including the Harbor District Plan, which aims to preserve industrial zoning to protect small manufacturers, and the City's Anti-Displacement Plan.

In Denver, the Mile High Connect partnership works to improve and preserve existing and build new affordable housing, encourages retail and supports small businesses, fosters libraries and other community assets, increases resident engagement in planning process, and makes transit more accessible and affordable to low-income populations (Lung-Amam et al. 2014).

Baltimore's community benefits agreement (CBA) was negotiated for the city's Red Line LRT system. The CBA includes a list of goals and strategies to generate jobs, develop economic opportunities, maintain affordable housing, and invest in impacted low-income communities. It was signed by 70 businesses and organizations in 2009. Though it's not legally binding, it is widely considered to be a success (Lung-Amam et al. 2014).

In Atlanta, an ordinance adopted in 2016 intends to compel the city, regional transportation authority MARTA, and Atlanta BeltLine Inc. to share plans and collaborate on their various transportation projects, including ensuring lower-income residents are hired to work on construction projects and that minority- or female-owned companies receive a significant portion of spending. Community Benefits Agreements specified in the ordinance are intended to guide a share of construction dollars to demographics that do not historically benefit from transportation infrastructure projects. However, the ordinance is advisory, not regulatory (Pendered 2016).

Generating Support from Higher Levels of Government

Several of our case studies, including Toronto, Montreal, and Arnhem-Nijmegen, had difficulty implementing TOD because of weak relationships or unstable funding from higher levels of government. TOD approaches must include multiple levels of government funding to be financially viable, yet we heard from our Dutch participants that their relationships with the national government were lacking. There are very few examples of successful generation of support from state or federal authorities; those cases that have been successful in this area (e.g., Naples) are quite unusual. This is a key challenge when trying to integrate equity considerations into TOD.

In Maryland, the Purple Line Corridor Coalition (PLCC) includes community officials, non-profits, and businesses across the proposed 16-mile (25 km) LRT line. PLCC was organized to fight for affordable housing, small business incubation, and the inclusion of historically under-represented communities in the planning process. One of the major achievements was the Purple Line Compact, a non-binding "statement of intent" that defines and supports goals for the corridor: preserving affordable housing, supporting small businesses, and connecting workers to jobs (Lung-Amam et al. 2014). Langley Park, a working-class, multicultural community along the proposed route, received special attention from the state: it was designated a Sustainable Community by the State of Maryland and was one of six in the county to participate in the Transforming Neighborhoods Initiative in 2012, which focuses on improving key indicators of community well-being, including addressing economic health, public safety, and educational challenges. The community was selected for a Department of Education Promise Neighborhood Planning Grant in 2014, which will improve the public school system, create a new bilingual high school, and launch an organization to increase and maintain access to affordable housing, jobs that pay a living wage, and locally owned businesses. Local organizations also received funding from the State Department of Labour, Licensing and Regulation to help small businesses adjust during the Purple Line construction period and a grant from the Department of Housing and Community Development to develop their own affordable housing strategies to prepare for the Purple Line.

Lung-Amam et al. (2014) also discussed how it can be difficult to preserve affordable housing when very little land or housing in the neighborhood is publicly held, owned by residents, or subsidized; 90 percent of small business owners did not own their buildings and had leases set to end within the next ten years. The neighborhood was also in an awkward situation because it is situated in an unincorporated area on the border of the municipal government and two county governments, each of which has a different level of interest in preserving and creating affordable housing and protecting small businesses. The authors maintain that "Ironically, however, Langley Park's awkward position on the border between the two counties with such vivid contrasts may serve it well" (2014, 13), because advocacy efforts and conversations about equity between the two counties might not otherwise have taken place. Maryland-National Capital Parks and Planning Commission, a regional organization encompassing both counties, provides a framework in which these collaborative planning solutions and strategies can emerge. Metro Washington Council of Governments (MWCOG) also submitted a grant application to Housing and Urban Development that would have funded a multi-stakeholder initiative for a comprehensive program of equitable TOD in the region. MWCOG's proposal was not funded, but its Region Forward 2040 transportation plan aims to encourage mixed-income development around transit and boost economic development in Prince George's County especially near fixed-route transit.

Conclusions

Integrating different cases into our comparison of international city-regions would invariably have led to different results, which is why this chapter focused on other factors that have been critical to successful TOD implementation in cases outside of ours. This rounds out our study, which was completed in 2014, when equitable TOD (eTOD) approaches were in their infancy. Since then, American cities have taken the lead in ensuring that community residents are receiving more of the benefits of TOD and fewer of the negative impacts, and community organizations have been instrumental in reshaping TOD planning processes to meet their own unique needs. Many of the tools and strategies presented in this chapter will not be new to readers; however, their use in redistributive planning processes is unique and will likely be new to transportation planners and engineers who have not yet considered the negative impacts of TOD on communities.

References

Bardaka, E., & Hersey, J. (2018, June 15). *Transit-oriented development is more transit-oriented when it's affordable housing.* TransitCenter. Accessed 11 Dec 2019.

Carlstedt, M., & Washburn, O. (2019, February 5). *$15m equitable transit-oriented development launched.* Building Indiana. https://www.buildingindiana.com/15m-equitable-transit-oriented-development-launched. Accessed 3 Apr 2019.

Chapple, K., & Loukaitou-Sideris, A. (2019). *Transit-oriented displacement or community dividends?* Cambridge, MA/London: MIT Press.

Chemtob, D. (2019, April 15). New rules for development near Charlotte transit may help affordable housing too. *Charlotte Observer.* https://www.charlotteobserver.com/news/business/biz-columns-blogs/development/article229273054.html. Accessed 29 July 2019.

City of Milwaukee. (2018). *TOD connects: Equitable growth through transit-oriented development. A neighbourhood plan for Walker's Point (Draft).* Milwaukee: City of Milwaukee.

City of Omaha. (2019a). *Transit oriented development (TOD) initiative.* City of Omaha website. https://urbanplanning.cityofomaha.org/master-plan/48-long-range/103-transit-oriented-development. Accessed 31 July 2019.

City of Omaha. (2019b). *Transit oriented development guide: Park Avenue ORBT station.* City of Omaha. Draft April 11, 2019.

City of Omaha. (2019c). *Transit oriented development guide: 35th/22rd Street ORBT station*. City of Omaha. Draft April 11, 2019.

Enterprise Community. (2018). *Denver Regional TOD Fund*. Enterprise Community website. https://www.enterprisecommunity.org/financing-and-development/community-loan-fund/denver-regional-tod-fund. Accessed 2 Dec 2018.

Espeseth, D. (2017). *Equitable TOD in Halifax: Exploring rental housing strategies and partnerships in transit-oriented developments* (Masters independent study project). Halifax: School of Planning, Dalhousie University.

Federal Transportation Administration. (2017). *Transit-oriented development technical assistance: Selected projects*. https://www.transit.dot.gov/funding/funding-finance-resources/transit-oriented-development/transit-oriented-development. Accessed 2 Nov 2018.

Grimshaw, J. (2019, January 29). *Chicago's latest ordinance expands eTOD*. Centre for Neighbourhood Technology blog. https://www.cnt.org/blog/chicago%E2%80%99s-latest-ordinance-expands-etod. Accessed 2 Feb 2019.

Grube-Cavers, A., & Patterson, Z. (2015). Urban rapid rail transit and gentrification in Canadian urban centres: A survival analysis approach. *Urban Studies, 52*(1), 178–194.

Jannene, J. (2018, October 23). *New development plans for streetcar*. Urban Milwaukee. https://urbanmilwaukee.com/2018/10/23/eyes-on-milwaukee-new-development-plans-for-streetcar/. Accessed 11 Dec 2018.

Jones, C. (2015). *Transit-oriented development and gentrification in Metro Vancouver's low-income SkyTrain corridor*. Toronto: Cities Centre, University of Toronto.

Lung-Amam, W., Pendall, R., Scott, M., & Knaap, E. (2014, October). The promise and challenge of equitable transit-oriented development in diverse suburbia. In *Paper presented at transit, development and Forme Urbaine: Washington et Paris symposium*. Washington: DC.

Mueller, E. J., Hilde, T. W., & Torrado, M. J. (2018). Methods for countering spatial inequality: Incorporating strategic opportunities for housing preservation into transit-oriented development planning. *Landscape and Urban Planning, 177*, 317–327.

Pendered, D. (2016, October 2). *Atlanta City Council back on track with social equity for transportation sales taxes*. Saporta report. https://saportareport.com/atlanta-city-council-back-track-social-equity-transportation-sales-taxes/. Accessed 11 Dec 2018.

Pollack, S., Gartsman, A., Benedict, A., & Wood, J. (2014). *Rating the performance of station areas for effective and equitable transit-oriented development*. Paper presented at Transportation Research Board annual meeting, Washington, DC, January 12–16, 2014.

Poverty and Race Research Action Council. (2015). *Equitable transit oriented development: Examining the progress and continued challenges of developing the affordable housing in opportunity and transit-rich neighbourhoods.* Civil Rights Research Report.

Renne, J. L., Tolford, T., Hamidi, S., & Ewing, R. (2016). The cost and affordability paradox of transit-oriented development: A comparison of housing and transportation costs across transit-oriented development, hybrid and transit-adjacent development station typologies. *Housing Policy Debate, 26*(4–5), 819–834. https://doi.org/10.1080/10511482.2016.1193038.

Sandoval, G. F. (2016). *Developing a model for transit oriented development in immigrant communities: A national study of equity and TOD.* Portland State University: Transportation Research and Education Center (TREC). https://doi.org/10.15760/trec.160

Sandoval, G. F. (2018). Planning the barrio: Ethnic identity and struggles over transit-oriented, development-induced gentrification. *Journal of Planning Education and Research.* https://doi.org/10.1177/0739456X1879371.

TransitCenter. (2018, February 2). *Affordable housing: A next frontier for transit?* TransitCenter. http://transitcenter.org/2018/02/06/a-next-frontier-for-transit/. Accessed 11 Dec 2018.

CHAPTER 5

Conclusion

Abstract TOD has long been considered a critical element in reshaping sprawling car-dependent urban regions, common in many countries, to denser rail-based regions. The concept has now evolved far beyond railway station areas to corridors and entire multi-modal networks designed around not only transit (heavy rail, LRT, BRT, and local bus) but walking and cycling as well. But TOD is not a magic bullet solution for metropolitan transportation problems: it has its challenges, such as displacement of local residents and businesses. Although TOD is still challenging to implement, we hope that our list of critical success factors and ranking scale will help contribute to successful TOD implementation in any region. TOD is a policy concept that can be used, as our Dutch colleagues suggested, as a story to unite people: a story that includes sustainable transportation, healthy populations, thriving local economies, coordinated strategic planning, and an engaged public.

Keywords Development • Cities • Regions • Implementation • Planning tools

Transit-oriented development has many potential benefits for cities and regions: high-density development along high-frequency fixed-route transit makes efficient use of infrastructure, facilitates higher transit ridership which can reduce congestion and air pollution, and can benefit local

R. Thomas, L. Bertolini, *Transit-Oriented Development*,
https://doi.org/10.1007/978-3-030-48470-5_5

economies through job creation during and after project construction. TOD has long been considered a critical element in reshaping sprawling car-dependent urban regions, common in many countries, to denser rail-based regions. The concept has now evolved far beyond railway station areas to corridors and entire multi-modal networks designed around not only transit (heavy rail, LRT, BRT, and local bus) but walking and cycling as well, which can have public health benefits.

But TOD is not a magic bullet solution for metropolitan transportation problems, and in fact it causes its own challenges. Overly complex development processes, financial and construction challenges, and negative impacts such as displacement of local residents and businesses are also characteristics of TOD. This is precisely why it is critical to learn from built examples, both those that are successful and those that are struggling to attract new businesses, residents, or jobs.

In this book, we have summarized an in-depth study we conducted at the University of Amsterdam (2012–2014). Our meta-analysis of 11 international case studies, including meta-matrix, rough set analysis, and policy transfer workshops, has been discussed at length in articles in several academic journals. In this volume, we were able to discuss TOD and its benefits and challenges much more thoroughly: in particular, Chaps. 1 and 4 refer to the latest literature and issues that we did not consider in the original study.

Cities and regions have introduced TOD plans and policies to support dense regional development along existing and future transit routes, introduced tools to support TOD in local zoning codes and by-laws, and written design guidelines for station areas and main streets. As the demand for dense, urban neighborhoods within walking distance of shops and services increases, there is a lot of opportunity for infill development in many transit-served cities and suburbs. In countries like the US, Canada, and Australia, with vast amounts of land for development, car-oriented land use patterns, and predominantly private-sector development (e.g., of housing), planning incentives like these are necessary. Even in European, Asian, and other densely built, rail-oriented countries, with stronger public-sector involvement, tools like FAR bonuses, tax incentives, decreased parking ratios, and density bonuses can make sites near transit infrastructure far more attractive to developers. There are many successful examples of TOD around the world, but how many cities and regions have been able to successfully shift the development pattern away from the car and toward fixed-route transit? How many have seen transportation mode shifts that are significant enough to decrease air pollution and congestion? How many have benefitted the local economy without displacing people?

Clearly, TOD is still challenging to implement. Particularly when planners, politicians, and civil servants do not have examples of policy solutions in their own countries, or where there is a lack of expertise in policy development, practices of learning from other countries have been established. While it can be tempting to copy policy ideas from places that are similar to our own political, social, and geographic context, it seems to be more effective to use these ideas as sources of learning and inspiration. Otherwise, we risk copying policy ideas incompletely or inappropriately. The Dutch situation illustrates that some parts of TOD as a policy concept (e.g., high-density, mixed-use development in station areas) have made it to the Netherlands, while necessary characteristics of the development process (e.g., strong actor relationships, willingness to experiment, public engagement) have not. In part, these omissions have contributed to persistent challenges in implementation: TOD is still seen as an infrastructure concept and has not reached a critical level of dissemination or discussion among land use planners, politicians, sociologists, economists, urbanists, or the general public.

In studying the 11 case city-regions, we hoped to develop a list of factors that would contribute to successful TOD implementation in any region in the world. While we certainly cannot be this bold, we believe that the list of 16 critical success factors can be used to understand the particular strengths and weaknesses in a region (Chap. 3, Fig. 3.2):

1. Policy consistency
2. Vision stability
3. Government support
4. Political stability (national)
5. Political stability (local)
6. Actor relationships
7. Regional land use-transportation body
8. Inter-municipal competition
9. Multidisciplinary implementation teams
10. Public participation
11. Public acceptance
12. Key visionaries
13. Site-specific planning tools

14. Regional-level TOD planning
15. Certainty for developers
16. Willingness to experiment

Using the five-point ranking scale (Chap. 3, Fig. 3.3), planners, engineers, developers, community members, and other stakeholders can evaluate their own region and collectively decide where to focus their efforts in the future. This process, along with informal meetings between the main actors and a concerted effort to engage the public and win public approval, will strengthen relationships and build a collaborative practice, benefiting any region.

Because our findings are directly related to our choice of case studies, including the derivation of critical success factors from the meta-matrix, we have given examples of persistent concerns with TOD from other contexts, particularly the US. American cities have become leaders in addressing some of the negative impacts of TOD, such as displacement of residents, disruption of local businesses during construction, and lack of jobs and services for local people in new developments. Not only do these efforts benefit transportation authorities, who rely heavily upon low-income, minority, youth, senior, and immigrant populations to keep ridership levels high, but many of the tools and strategies used in equitable TOD (eTOD) actually address persistent barriers such as high land costs near transit, public opposition, and zoning/regulatory barriers to high-density mixed-use development.

Transportation authorities and local planning organizations can use several tools, like the Housing + Transportation Index in the US, to learn which neighborhoods are more vulnerable to the loss of affordable housing and then buy or facilitate the purchase of sites in these areas for future development of housing and community services tailored to the local low-income populations. Examples from cities as diverse as Charlotte, Los Angeles, and Milwaukee illustrate the planning regulations such as TOD zones and overlay zones, requirements for affordable housing in station area developments, and density bonuses that require inclusion of affordable units. In cities like San Diego, Oakland, Denver, Baltimore, and Atlanta, collaborative practices have been developed to facilitate community benefits agreements, community-led development, and neighborhood plans that reflect the needs of ethnic and low-income populations. It may even be possible to use equity considerations to garner the support of higher levels of government for a vulnerable neighborhood, as happened in Maryland, or to be so successful at eTOD that the area becomes a national success story, as happened in Fruitvale (Oakland).

Transit-oriented development is much more than building high-rises at railway stations. It is a policy concept that can be used, as our Dutch colleagues suggested, as a story to unite people: a story that includes sustainable transportation, healthy populations, thriving local economies, coordinated strategic planning, and an engaged public. We hope that readers can use this book as a guide to write their own story in a way that acknowledges the uniqueness of their own context.

Index

R. Thomas, L. Bertolini, *Transit-Oriented Development*,
https://doi.org/10.1007/978-3-030-48470-5

The manufacturer's authorised representative in the EU is Springer Nature Customer Service Centre GmbH, Europaplatz 3, 69115 Heidelberg, Germany. If you have any concerns regarding our products, please contact ProductSafety@springernature.com

Printed and bound by CPI Group (UK) Ltd, Croydon, CR0 4YY
30/09/2026
02225609-0002